FROM CELLS
to
ORGANISMS

SHERRIE L. LYONS

FROM CELLS
to
ORGANISMS

RE-ENVISIONING CELL THEORY

UNIVERSITY OF TORONTO PRESS
Toronto Buffalo London

© University of Toronto Press 2020
Toronto Buffalo London
utorontopress.com

ISBN 978-1-4426-3510-4 (cloth)   ISBN 978-1-4426-3511-1 (EPUB)
ISBN 978-1-4426-3509-8 (paper)   ISBN 978-1-4426-3512-8 (PDF)

Library and Archives Canada Cataloguing in Publication

Title: From cells to organisms : re-envisioning cell theory / Sherrie L. Lyons.
Names: Lyons, Sherrie Lynne, 1947– author.
Description: Includes bibliographical references and index.
Identifiers: Canadiana (print) 20200210203 | Canadiana (ebook) 20200210211 | ISBN 9781442635104 (cloth) | ISBN 9781442635098 (paper) | ISBN 9781442635111 (EPUB) | ISBN 9781442635128 (PDF)
Subjects: LCSH: Cells. | LCSH: Cytology. | LCSH: Organisms. | LCSH: Developmental biology. | LCSH: Heredity.
Classification: LCC QH581.2 .L96 2020 | DDC 571.6 – dc23

We welcome comments and suggestions regarding any aspect of our publications – please feel free to contact us at news@utorontopress.com or visit us at utorontopress.com.

Every effort has been made to contact copyright holders; in the event of an error or omission, please notify the publisher.

University of Toronto Press acknowledges the financial assistance to its publishing program of the Canada Council for the Arts and the Ontario Arts Council, an agency of the Government of Ontario.

Cover illustration: Adolf Giltsch, *Thalamophora. – Kammerlinge*, 1904. Library of Congress, LC-DIG-ds-07540.

Canada Council   Conseil des Arts
for the Arts   du Canada

ONTARIO ARTS COUNCIL
CONSEIL DES ARTS DE L'ONTARIO
an Ontario government agency
un organisme du gouvernement de l'Ontario

Funded by the   Financé par le
Government   gouvernement
of Canada   du Canada | Canada

Dedicated to the memory of
Daniel Mazia

Contents

Illustrations

Preface

The genesis of this book began 25 years ago when I proposed an idea for my PhD dissertation at the University of Chicago. I wanted to look at the history of modern cell biology focusing on the work of Daniel Mazia. I was privileged to have him as a professor when I was an undergraduate at Berkeley and did an independent study with him my senior year. My advisor, Robert Richards, was a Darwin scholar who greeted my idea with lukewarm enthusiasm. The Committee on Evolutionary Biology at Chicago was a dynamic group and had a particularly outstanding group of graduate students at the time. Most of its members supported Stephen Gould's and Niles Eldredge's theory of punctuated equilibria that was creating a storm of controversy in the wider evolutionary biology community at the time. Gould and Eldredge began their classic paper with Thomas Huxley's caution to Darwin on burdening his theory unnecessarily with gradualism. For my dissertation I decided to trace the gradualist-saltationalist debates or, in the words of David Raup, the "creeps and jerks" in the history of evolution, beginning with Darwin who was a creep and Huxley who was a jerk. However, Huxley was a multifaceted person with a research program quite distinct from Darwin's, which raised several issues that were problematic for Darwin's theory. Because there was more than enough material for a dissertation on just Huxley, I have no regrets that I ended up doing my dissertation on him.

Huxley still fascinates me. In fact, many of the issues he raised as a developmental morphologist are only finally being addressed now. I stayed in contact somewhat irregularly with Mazia over the years as my interest in cell biology never disappeared. Many books have been and continue to be written on both the history of evolution and the history of genetics, but relatively few on cell biology. I thought I could tell a story that would highlight the work of both Huxley and Mazia.

Huxley, of course, was well known and highly respected, as was Mazia among their own colleagues. Yet I think I have become a historian of scientists who have not gotten the recognition they deserve. Huxley, like many of Darwin's cohorts, has been in Darwin's shadow in terms of his own scientific research. Likewise, researchers in modern cell biology have been somewhat in the shadow of molecular biology and the obsession with DNA, in both the scientific community and the public at large. Both Huxley and Mazia were outstanding teachers, profound thinkers, and imaginative and brilliant writers. In the course of writing this book, I have also discovered many other scientists whose contributions have been underappreciated. I hope in telling this larger story about cell theory that readers may come to appreciate and learn about some of the scientists whose work has contributed significantly to our understanding of life.

Sherrie L. Lyons
August 2019

Acknowledgments

Many people have been quite helpful in the writing of this book. To begin at the beginning, I would probably never even have thought of writing it if it wasn't for Kris Gies, who engaged me in a conversation at the UTP book display at an HSS meeting. He asked me if I could write a book about anything what would it be? I had just given a talk about Daniel Mazia and cell biology, and as we talked he thought I had some interesting ideas. He encouraged me to submit a proposal to his editor, Natalie Fingerhut. That same evening over dinner, Brian Hall helped me immensely in clarifying my thoughts and sketching out ideas that should be included in a formal book proposal. Over the years he has continued to be a great help, and I have benefited immensely from his own writings. My editor, Natalie Fingerhut, also helped shape the final proposal before she sent it out to reviewers. The reviewers were generally quite positive, and all had wonderful suggestions that encouraged me to expand the topics I covered and to add more actual science. Ms. Fingerhut totally supported me and backed me through some difficult times as I continued to reshape the manuscript. In addition, I am extremely grateful to Alex Grieves and Eileen Eckert for their assistance in bringing the manuscript to fruition. Marsha Richmond had many helpful suggestions and also sent me various articles that I had trouble accessing. Her paper on Huxley's critical review of cell theory was essential to my own analysis of Huxley. Conversations with Gar Allen and Jane Maienschein were also a help in discussing the various issues that were important to the story I was trying to tell. František Baluška's work made me think there was an interesting story that could link Thomas Huxley's ideas to those of Mazia's. In addition to being a cell biologist, he has a very good understanding of the historical characters in my story and suggested several people I should include as well as pointing out various errors while reading the entire manuscript. David Epel provided several key articles to read, and he read the chapter on Mazia. In addition, he provided additional information about Mazia in a long phone interview I had with him. Douglas Allchin has also been a tremendous help in thinking about science as a process, and about how the "standard" histories of various scientific episodes are often quite misleading. He critically read the manuscript before its final submission. Elaine Ostander replied to my questions about developmental plasticity in dogs and cats. Conversations with Sam Bowser and a tour of the Wadsworth Center, including one of the first electron microscopes, reinforced the importance of microscopy in the history of cell biology. Nikki Everts, Stuart Fryer, and Josephine Carenza pointed out places in the manuscript that seemed too technical or weren't clear. Josephine had several suggestions about how to make the manuscript more effective in helping people understand and retain what they were reading. Robyn Reed at Schaffer Library of Union College helped in providing the images. Last, but certainly not least, is Scott Gilbert, whose enthusiasm and superb knowledge of virtually all aspects of this manuscript has been invaluable. He critically read an earlier version of the entire manuscript, pointing out various factual errors, making important and also sometimes very entertaining suggestions. As always, however, any errors are my own. I am incredibly grateful that I had the opportunity to be a student of Mazia's. I hope this book will bring the work of this remarkable and creative scientist to a larger audience. Finally, numerous friends and my two wonderful children, Cassandra and Grahame, not only encouraged me, but also remind me what is truly important to value in this life.

Introduction

Biology has very few laws or unifying theories compared to physics. In fact, some people claim that there are only two: the cell theory of Matthias Schleiden and Theodor Schwann, and Charles Darwin's theory of **evolution**. These two core ideas were articulated around the same time that biology emerged as a distinct field in its own right, separate from anatomy and physiology and also from natural history. Biology has come a long way since then, building on these two powerful ideas. While much has been written about the history and controversies surrounding Darwinian theory, far less has been written about the history of the cell. General histories of biology provide good accounts of the many different ideas concerning the organization of life pre–cell theory, but they have been far less critical about the controversies surrounding the cell theory as articulated by Schleiden and Schwann in 1839. The theory consists of three core ideas:

1 All life is made up of **cells**.
2 Cells are the smallest independent unit of life.
3 All cells arise from preexisting cells.

The third idea, which was critical to the widespread acceptance of the theory, is attributed to Rudolf Virchow in 1855 but, as will be explored in greater depth in chapter 3, credit should really go to Robert Remak. This is one of many examples that contradict the standard canonical accounts of cell theory. Furthermore, just like evolution, cell theory was not immediately accepted and aspects of it have remained controversial to the present day. To say that the cell is the basic unit of life is a rather abstract concept and, as Andrew Reynolds has persuasively argued, the whole history of cell theory has relied heavily on the use of metaphors and analogies that are more descriptive and reflect differing ideas at a particular moment in history.[1] This standard definition of cell theory, however, provides the backdrop for the historical critique embedded in the text as I invite the reader to "re-envision" cell theory.

These introductory remarks provide a brief overview of the topics that will be explored in the following chapters. Examining the reception and history of cell theory provides a wonderful opportunity to elucidate important themes and controversies in biology. These include debates over **vitalism**, the meaning of biological individuality, and how **development** occurred: **preformation** or **epigenesis**. The theory of preformation claimed that the organism was fully formed at conception and that development was just a matter of growth. According to epigenesis, the embryo develops or unfolds by successive **differentiation** from an original undifferentiated structure. Underlying these questions was also a philosophical debate: Would a structural or functional approach yield the greatest insights into understanding biological phenomena?

A key player in many of these controversies was Thomas Henry Huxley (1825–95) (see figure 0.1). Best known as a popularizer and defender of evolution, he has been somewhat in Darwin's shadow. However, he was a first-rate scientist in his own right, a developmental morphologist with a research program quite distinct from natural

---

1 Reynolds, *The Third Lens*, 2.

selection and evolution. Huxley was interested in the fundamental question of how form comes to be. How does an egg become a giraffe or a redwood? Huxley had no use for vitalistic theories about the origin of life. One would think he would have strongly supported cell theory, which helped end the debates over vitalism – the idea that life is animated by some immaterial attribute or principle that is distinct from either purely chemical or physical forces. However, Huxley voiced criticisms of the theory, claiming that cells were not anatomically independent but were interconnected, creating larger assemblages. Therefore, cells could not be the elementary units of life. Huxley's objections also highlight a tension between cellular and organis-

0.1 Thomas Henry Huxley

mal concepts of life. His perspective contributed to an epigenetic tradition in the understanding of development and emphasized what was inside the cell, irrespective of cell boundaries.[2]

What constitutes a biological individual? Much of Huxley's research shaped how he answered this question.[3] In spite of being a morphologist, Huxley's emphasis was on how the cell maintained life, that is, the physiological function, which was the basis of the cell's vitality rather than its form. At the same time, he also thought that the whole organism guided and shaped how development proceeded. Huxley's views were a combination of what has been characterized as bottom-up and top-down approaches. One had to study the interactions of molecules and proceed upward to the level of cells, tissues, and the organism. At the same time, he thought such a research program was inadequate to fully understand development. One also had to start from the "top," the organism, and investigate how it influenced development. His views

---

2 Richmond, "T.H. Huxley's Criticism of German Cell Theory," 247–89.
3 Huxley, "Upon Animal Individuality," 172–77.

will provide the scaffolding for exploring the ongoing controversies that cell theory played in understanding this amazing phenomenon that we call life. Many of Huxley's ideas have found resonance with the concerns of present-day researchers working in evolutionary, developmental, and cell biology. Some biologists have suggested that cell theory may be in need of revision. Indeed, some of the most intriguing research today in cell biology reflects the kinds of issues that interested Huxley.

Prior to the twentieth century, **heredity** and development were considered together as a single field, complementary aspects of the fundamental problem of **generation**. At the turn of that century they became separate disciplines as the new discipline of genetics emerged. With this separation, heredity became much more narrowly defined, resulting in significant progress in elucidating the laws of heredity. But it also meant that certain fundamental questions were set aside. In particular, how does cell differentiation occur? The spectacular success of genetic investigations, culminating with the elucidation of the structure of DNA and how it carried out its function as the hereditary material, has resulted in genetics dominating the thinking in understanding a variety of biological phenomena – from evolution to cancer. **DNA** has been regarded as the "master molecule" that contains the blueprint for life. It is often described as a self-replicating molecule, but it accomplishes its hereditary function within the context of a host of other molecules as well as particular structures in the cell. Furthermore, understanding development has proven to be a much tougher problem than understanding heredity as defined in this more narrow sense.

When it was discovered that the cell **nucleus** contained the hereditary material, many biologists thought that it must also control development, but not everyone agreed. In fact, initially not everyone even accepted that the nucleus contained the heredity determinants; instead they argued that the **cytoplasm** was responsible for guiding development. The old debate between preformation and epigenesis became reformulated but still was not resolved. Chromosomes were in a sense preformed units that were inherited from the past. Yet as development proceeded, new structures were formed. Identifying the chromosomes as the carriers of the hereditary information still did not address the fundamental problem of how a single cell eventually gave rise to many different cell types. Research clearly demonstrated that the egg was highly organized; the cytoplasm was not just a homogeneous blob. The question remained: did the cytoplasm or the nucleus control development? Were the hereditary units divided up unequally in the daughter cells as development proceeded? How cell differentiation occurred remained an open question.

Examining the details of the split between heredity and development reveals that, nevertheless, the origin of the **gene** theory of inheritance was heavily informed by ideas from **embryology**.[4] The two needed to be brought back together to fully understand development. Yet attempts to truly reintegrate the two different disciplines met with limited success, which also had ramifications for evolutionary theory. Population genetics monopolized evolutionary theorizing for most of the twentieth century. The standard definition of evolution was "a change in gene frequencies." But such a definition is the *result* of evolution and tells us very little about the *process* of evolution.

Development was essentially left out of the the Modern Synthesis of the 1940s, which vindicated natural selection as the primary **mechanism** of evolutionary change. However, as Conrad Waddington (1905–75) wrote in 1975, "The evolution of organisms must really be regarded as the evolution of developmental systems."[5] But development cannot just be explained by natural selection causing changes in gene frequencies.

Due to the complexity of developmental processes, unraveling how a cell becomes a heterogeneous multicelled organism has not been characterized by the dramatic breakthroughs recounted in histories of heredity. Yet the story is a fascinating one, full of controversy and remarkable experiments. Many of these experiments depended on the continual advances in microscopy that revealed truly astonishing results. In addition to the many scientists, the microscope is an essential actor in the story I am telling. Daniel Mazia (1912–96), in particular, emphasized the importance of microscopy, maintaining that the major advances in our understanding of the cell have been made by seeing. He was convinced that revealing the underlying structure of the cell would provide insight into development (see figure 0.2).

To understand development, you must first understand, in the words of Mazia, how "one cell becomes two." Mazia can be considered one of

---

4 See Gilbert, "Embryological Origins," 307–51.
5 Waddington, *The Evolution of an Evolutionist*, 7.

0.2 Daniel Mazia. Courtesy of Harold A. Miller Library, Hopkins Marine
Station of Stanford University.

the founders of modern cell biology. He is best known for his work on
mitosis and cell reproduction. Although much of his career was devoted
to illuminating the process of how chromosomes precisely duplicate
themselves, Mazia was not a geneticist. He made clear that events in the
mitotic cycle ran through the entire **cell cycle**. It was not just what was observed during mitosis that determined chromosome duplication and cell
division. Furthermore, other researchers had demonstrated that tissues
and organs formed as the result of multiple molecular interactions, many
of which were not under the direct control of the genome. Mazia came
to think that something fundamental was missing in our understanding
of the cell. Late in life and until his death in 1996, he was working on
what he called the **cell body**. Its structure was smaller than the cell, but
had all the basic attributes of a living entity. Mazia's research also pointed
out certain inadequacies of cell theory. His research played an important
role, albeit indirectly, in a larger story that was unfolding. Heredity and
embryology needed to be reunited to fully understand development and
to deepen our knowledge of evolutionary processes.

How is form maintained; how does it change? Only at the end of
the twentieth century were genetics and development being brought
back together with the rise of a new research program, evolutionary

developmental biology or "evo-devo." Evo-devo finally began to address the question that most interested Huxley and was elaborated by Waddington. It is a variant of the age-old question, "Which came first, the chicken or the egg?" Cell theory states that the egg gives rise to the chicken, but one could also claim the egg is just the way a chicken makes another chicken. The organism is informing the actions of the cell, just as the actions of the cell determine the action of the organism. To fully understand development, the organism must also be considered as a biological unit of analysis, and cell-to-cell communication plays a critical role in the link between the **genotype** and **phenotype** of all organisms. For most higher plants, cell theory has always been problematic, since many of the individual cells have perforated walls that allow cytoplasm and nutrients as well as water to flow between them. Not only do plant cells communicate within the organism, but they also send messages to other plants, alerting their neighbors of attacks by herbivores and pathogens and even cautioning them of a coming drought.[6] Animals move and have an enormous behavioral repertoire, which also means that the communication between cells plays a significant role in determining their phenotype. The organism is more than the sum of its parts.

This history goes full circle and ends where it began with the problem of generation. Heredity and development are inextricably interconnected, and the advances in understanding heredity owe much to its developmental origins. Development begins with a single cell and, thus, the history of cell theory provides a window to view the progress made in understanding this complex process. The history of development also shows that investigations in understanding the cell proceeded in two very different directions, revealing a tension between **reductionism** and a more holistic approach to understanding life.

As powerful as the reductive mechanistic approach has been, particularly in the field of genetics, the techniques that have been so successful in understanding heredity have limitations for understanding development. Biochemists ground up and dissected the cell in an attempt to understand the biochemical pathways that were responsible for the processes that make life possible. Cell biologists, making use of more and more powerful microscopes, saw that the cell was highly

6 Cossins, "Plant Talk," 37–43.

structured, containing many different organelles besides the nucleus. More evidence unequivocally showed that the whole organism influenced how development proceeded. This story reveals that neither a strictly biochemical approach nor a morphological one is adequate to understanding how development proceeds. The form versus function debates that were so prominent in the nineteenth century are no longer useful. An integrated approach is needed to address the kinds of concerns that both Huxley and Mazia raised. While today no one disputes that organisms are made up of cells or that they are interacting with one another, cell theory may nevertheless need to be modified to fully understand how a cell becomes an organism, surviving, reproducing, and evolving in an ecosystem.

The controversies surrounding cell theory provide an important case study that explores such questions as how and why some topics become subjects of investigation at a particular moment in history. Debates over vitalism vs. mechanism that characterized much of the first part of the twentieth century were slowly leading to a synthetic **organicism** that was going to have far-reaching implications for the kinds of experimental programs designed to understand the structure of organisms.[7] Tracing the history of cell theory illustrates that the "facts" do not speak for themselves, and observations always need to be interpreted. Yet this history also reveals that scientists guided by different philosophical underpinnings and making use of a variety of different experimental techniques are moving us to an ever-increasing understanding of the fantastic phenomenon we call life.

---

7 Haraway, *Crystals, Fabrics, and Fields*, 6.

CHAPTER ONE

Microscopes and the Discovery of the Cell

There was such an enormous number of living animalcules here, that I imagined I could see a good one thousand of 'em in a quantity of this material that was no bigger than a hundredth part of a sand-grain.

Anton van Leeuwenhoek describing what he saw from a scraping of plaque from his teeth, 1683

WHAT IS LIFE?

What is life? This is a question that has occupied humans since we possessed brains large enough to contemplate such problems. Our earliest ancestors believed that all objects were imbued with their own unique spirit that gave them their particular characteristics, whether it was a rock, a river, a cat, or a rose. Some people even today have such thoughts, but beginning with the Greeks (if not earlier), people felt that there was a difference between an inanimate object such as a rug and the cat that was curled up on it. This vitalistic force that animated living creatures was the *psyche* (the Greek word often translated as soul, but having little relationship to the Judeo-Christian conception of soul), and it was no longer present when an organism died. In Aristotle's extensive investigation of a wide range of biological organisms, he described their structure and tried to understand their development and means of reproduction. In doing so he hoped to find answers to a more fundamental question: what is the nature of this vital force that allows an egg to grow, differentiate, mature, reproduce, but leaves when the organism dies? Living organisms are characterized by constant change. Inanimate objects change, but not in such a dramatic fashion, and they do not reproduce. Is this life force essentially the same for all living beings, and if so, how is it able to manifest itself in such different creatures – fish, flowers, trees, dogs, and humans? How is life maintained? What is the cause of such diversity, not only the variation seen within species, but also between species? What is responsible for like producing like or, in modern-day parlance, what is the hereditary material and how is it transferred from generation to generation? It is going to take more than 2000 years before these questions are answered in detail, and the most basic one – how does life differ from nonlife – still defies a complete explanation. However, the discovery of the cell and development of cell theory played an absolutely critical role in answering these most basic questions. Cell theory is part of a tradition that has been called *biological atomism* – the idea that all life is composed of elementary and indivisible units.[1] A living organism is the product of all the interactions and activities of its constituent parts, but each of those parts has the attributes of what makes something "alive." The history of cell theory reveals that what constitutes this primary element has been a source of controversy, with various theories suggesting units that are either smaller or larger than the cell.

As is often the case in science, significant breakthroughs are dependent on advances in technology, particularly ones that extend the abilities of our own senses. The telescope allowed Galileo to observe the moons of Jupiter, objects that were very large, but also extremely far away. The microscope allowed us to observe the very small up close, and revealed an unseen world brimming with life. We now know that microorganisms account for 99 percent of all life forms and that there are as many bacteria in us as our own cells. Yet it was not until the 1600s, with the invention of magnifying lenses, that we were able to see these organisms. A single double concave magnifying lens or simple microscope seems to have been around since the 1500s, which magnified objects

---

1 Nicholson, "Biological Atomism and Cell Theory," 202–11.

a few diameters, allowing one to see fascinating details of structures of various small objects, both living and nonliving. However, considerable improvements were needed to reveal the world of microorganisms. Those improvements were made by Anton van Leeuwenhoek (1632–1723), a Dutch shopkeeper from Delft.

ANTON VAN LEEUWENHOEK
AND HIS LITTLE ANIMALCULES

Anton van Leeuwenhoek was born into a Delft family that was part of the prosperous middle class of artisans, brewers, and minor public officials in what is considered the Golden Age of the Dutch Republic. In 1648 he was apprenticed to a cloth merchant in Amsterdam, but he returned to Delft in 1654 to begin his career as a shopkeeper. In 1660 he began a new career as a civil servant, first as an usher to the alderman of Delft, and then as a surveyor to the court of Holland. In 1677 he was made chief warden of Delft. Because of his mathematical skills, in 1679 he was made wine gauger, the inspector of weights and measures. The income from these various offices made him financially secure; in addition, the municipality granted him a pension in gratitude for his considerable scientific achievements.

Leeuwenhoek's scientific career can be considered to have started about 1671 when he constructed his first simple microscope. He got the idea from drapers, who used magnifying glasses to count threads and inspect the quality of cloth. From a glob of glass, he ground by hand a tiny lens and clamped it between two perforated metal plates. He then added a specimen holder that revolved in three planes. Secretive about his techniques, he made lenses of increasing quality, and ground about 550 lenses in his lifetime. Based on his drawings, the best of them must have achieved a magnifying quality of about 500 with a resolution of 1.0 μ (1 micron = 1×10^{-6} meters). To his lens-grinding he brought excellent eyesight, mathematical rigor, great patience, and extreme manual dexterity. With his sharp practical intellect, his new invention enabled him to explore virtually all areas of natural science for 50 years. His achievements are all the more remarkable since he never attended university and he worked in relative isolation most of his life. However, he read a great deal, mainly Dutch authors but also standard works in translation. In addition, he learned much from illustrations of books that were in foreign languages.

In 1673, the secretary of the Royal Society of London, Henry Oldenberg, initiated a correspondence with Leeuwenhoek. This began an exchange that Leeuwenhoek would continue with various members of the Society for 50 years, detailing his observations and discoveries. In 1680 he was made a Fellow of the Royal Society, a very high honor indeed; its members included such notables as Robert Boyle, William Cavendish, Robert Hooke, Isaac Newton, and Christopher Wren. Many of his letters would appear in the *Philosophical Transactions of the Royal Society*. Perhaps the most famous one was dated 7 September 1674, in which he describes his discovery of microorganisms:

> About two hours distant from Delft there is a lake with marshy or boggy places on the bottom. Its water is very clear in the winter, but in the summer it becomes whitish and there are then little green clouds floating in it which, according to the country folk, are caused by the dew. Passing the lake recently, and seeing the water as just described, I took a little vial of it. When I examined it the next day, I found earthly particles floating in it and some green streaks, spirally wound and orderly arranged. The whole circumference of each of these streaks was about the thickness of the hair on one's head [the streaks were the green alga *Spirogyra*]. Among these were also very many little animalcules, some of which were round-ish [protozoans]. Others, a bit bigger, were oval and I saw two little legs near at the head and two little fins at the hind end of the body [rotifers]. Others were somewhat longer than an oval and these were moved very slowly and were few in number [probably ciliates]. The motion of most of these animalcules in the water was so swift and so varied – upward, downward, and round about – that 'twas wonderful to see.[2]

Many members of the Society could not believe that such tiny "animalcules" could actually exist. However, by comparing their diameters to objects that could be directly measured and with detailed

2 Quoted in Dobell, *Antony van Leeuwenhoek*, 109–10.

calculations, he rebutted various objections. He developed a practical scale of micrometry for this previously invisible world, using as standards a coarse grain of sand, the width of a hair from his beard, and a red blood cell. He estimated that 27,000,000 of his animalcules could be fit into a grain of sand and that a cubic inch of soil could hold 13,824,000,000,000 of them. In addition, he sent testimonials and affidavits from various jurists, ministers, lawyers, and medical men, who vouched for the reliability of his results. Nevertheless, without his superior lenses, members of the Society could not duplicate his results and they sent a delegation to Delft. As a result of their rave report confirming Leeuwenhoek's observations, notables from Anne (the future Queen of England) to Tsar Pyotr I of Russia wanted a demonstration of these fantastic marvels.

As exciting and amazing as the discoveries of the animalcules were, Leeuwenhoek's work had even more profound implications. He used the microscope to further investigations concerning the most basic questions that continued to occupy the life sciences for the next several hundred years. Although the Royal Society was to play a pivotal role in establishing the role of experiment in generating knowledge, the world of seventeenth-century science was often characterized by wild speculation. Leeuwenhoek made a sharp distinction between his empirical observations and what he thought those observations actually meant. Nevertheless, his investigations were guided by two important ideas. He thought, first, that the inorganic and **organic** world were made up of the same basic building blocks and, second, that living organisms shared similar forms and functions. In order to support these assumptions, he investigated as many different organisms as possible and tried to generalize his findings. He drew analogies between animal and plant structures that sometimes allowed him to overcome difficulties in interpreting those microscopic structures. He discovered and accurately described red blood cells in a variety of different organisms, including fish, pigs, birds, and humans, as well as describing the blood capillaries. He made a distinction between the blood and the lymphatic capillaries that contained a "white fluid, like milk." He observed that yeast consisted of individual plant-like organisms. Everything was a source of his investigations. In feces he found a variety of different animalcules, bacteria as well as **protozoa**. The tartar on his teeth was also full of animalcules. He discovered that "fleas have fleas" and was the first to observe that the hundreds of facets making up a fly's eye were each an individual eye with its own separate lens. He was also the first to observe spermatozoa in the semen of various different species and referred to them as sperm animals.

To return to the fundamental questions of what is life and where does it come from, Leeuwenhoek hoped his investigations would disprove the still prevailing Aristotelian theory of **spontaneous generation**. Aristotle held that certain tiny animals such as insects and intestinal worms were spontaneously generated from the **putrefaction** of organic matter. In 1668 Francesco Redi (1626–97) thought he had successfully disproved the theory, at least in regard to the generation of maggots. In two jars he placed some meat, but one he covered with cloth while the other had no cloth over it. After a few days, the open one had maggots while the covered one did not. However, there were flies on it, and in other experiments Redi showed that the maggots came from fly eggs. Jan Swammerdam (1637–80) confirmed these experiments. Using a single-lens microscope he investigated the anatomy and reproduction of various insects. Leeuwenhoek furthered these investigations by describing the complex structure of mites, lice, and fleas and by his careful observations of their reproductive and life cycles. Nevertheless, the debate over spontaneous generation would not be settled definitively for another 200 years, as various researchers designed other experiments that seemed to demonstrate that some organisms were spontaneously generated from inorganic matter. It was not until the middle of the nineteenth century and the advent of cell theory, which played a definitive role, that this controversy was finally settled. Cell theory also played a prominent part in another controversy that had its roots in seventeenth- and eighteenth-century science: whether development occurred by epigenesis or preformation.

Epigenesis or Preformation

Regardless of whether some life forms were spontaneously generated or not, it had been recognized that, for higher organisms, life came from inside the mother. Aristotle had studied the development of chick embryos along with many other species and concluded that development began as a result of the union of the fluids of the male and female. For Aristotle, form gradually emerged from unformed material, the various organs appearing as development proceeded. This put Aristotle firmly in the epigenetic tradition, but, more importantly for this discussion, **epigenetics** at this point was also associated with vitalism. There was something other than physical or chemical forces that animated material objects and gave them the quality of life. For Aristotle, the menstrual blood only provided something "that out of which it generates." It was the male's contribution that animated this lifeless unformed material and made possible development. Such ideas guided investigations for well over a thousand years. Nevertheless, a debate arose. While the prevailing idea was that vapors arising out of the seminal fluid contributed a vitalistic principle that animated a lifeless egg, others thought it was the egg that contained the vital force. The most prominent proponent of this theory was William Harvey (1578–1657), best known for his demonstration of the circulation of the blood. In 1651 he claimed that the ova was the source of all life originating in the ovary of Eve.

However, ovism also raised another issue. According to the "spermists," the sperm was providing some vitalistic force that made it possible for new structures to be created in the developing embryo. **Ovism** was the first conceptual model of the idea of preformation, that an organism is formed by the unfolding and growth of already preformed parts. Epigenesis at that time was associated with vitalism because it suggested some vital force or intelligence was needed to organize the parental fluids into some recognizable structure.

The philosophical roots of preformation lie with René Descartes (1596–1650), who introduced the concept of materialism. For Descartes the world was mechanistic and deterministic. No intangible or vitalistic force was needed to create life. Although not explicitly stated, preformation seemed to be an argument against both epigenesis and spontaneous generation. Swammerdam also contributed to the concept of preformation, demonstrating that the rudiments of the adult structures such as legs and wings could be observed in the larval stages of various insects such as mayflies and butterflies. Marcellus Malphigi's (1664–96) research likewise supported preformation. He also worked on insects, but he is best known for his detailed illustrations of the various stages of chick embryos. Looking through the microscope, he was able to observe tiny, fully formed organs that needed only to unfold and grow to become a fully formed chick.

Later philosophers and naturalists suggested that these fully formed parts must exist from the beginning. The theory of preformation would have its greatest prominence in the early 1700s with the growth of the concept of *emboîtement*. The Catholic priest and philosopher Nicolas Malebranche (1638–1715) claimed that every life form on earth was made at the moment of creation. Future members of the species existed in the ovary of the first female of that species, like nesting Russian dolls, each one inside the other. The male only supplied seminal fluid to initiate growth of the already fully preformed embryo. This idea was challenged by Leeuwenhoek and Nicholas Hartsoeker (1656–1725), both of whom had discovered sperm. Rather than little Russian dolls inside the ovary, Hartsoeker suggested that in the head of each sperm was a tiny **homunculus** or "little man." His 1694 drawing of a curled-up infant human inside a sperm head became the exemplary image of the idea of preformation. Since Leeuwenhoek saw moving sperm and he associated motility with life, this put him firmly in the spermist camp. It is not known whether he actually postulated that the sperm penetrated the egg, since such a process couldn't be observed. In fact, it was not until 1826 that Karl Ernst von Baer was able to observe a mammalian egg. The ovist-animalculist controversy continued until 1875 when Oscar Hertwig demonstrated that **fertilization** was the result of the fusion of the nuclei from the sperm and the egg, and thus both were involved in creating new life.

To a modern reader, the idea that the first female of a species contained all the future individuals of that species or that a tiny homunculus sat inside the head of a sperm may seem patently ridiculous. But preformation was an important step in putting to rest the idea of spontaneous generation. In actuality, with its materialist underpinnings, it was the beginning of a line of research that undermined the various vitalistic theories that continued to abound. Yet both theories had serious shortcomings in trying to understand cell differentiation. As crude as their observations were, epigenesists' investigations showed that new structures did appear. Yet how did epigenesis account for development of new structures out of structureless material, form from the formless? Preformationists avoided this difficult problem of differentiation by claiming that the structures were there right from the beginning.

Yet it had problems as well. If the egg of a horse contained a preformed horse, how did one explain the mule when a horse mated with a donkey? How did preformation account for offspring that were intermediate when two varieties of plants were crossed? Furthermore, in the 1700s and 1800s, evidence was accumulating that some animals had amazing abilities of **regeneration**. Strict preformation would not allow for that process. At the same time, since offspring do resemble their parents, this suggests that something is conserved and passed on from generation to generation. Rabbits do not give rise to cats. Thus both theories accounted for some observations, but not others. How does development occur? The fundamental problem of embryology was indeed a thorny one, and one that has not been entirely solved even today. One would think, however, that the discovery of the cell in cork by Robert Hooke (who was working at the same time as these debates) would have settled these various questions, but the story is far more complicated and also far more interesting.

ROBERT HOOKE'S CELL

Robert Hooke (1635–1703) was truly a Renaissance man, making important contributions in architecture, astronomy, biology, chemistry, physics, map making, and surveying, as well as designing and building various scientific instruments. Hooke was born on the Isle of Wight in 1635. A frail child, he spent a large part of his school years at home. His clergyman father was impressed with his fine drawing skills and that he was able to help him work on various instruments such as clocks. Hooke's father died in 1648 and left his son £100, which was a considerable sum of money at the time. The 13-year-old Robert went to London to be educated at Westminster School, where he learned Greek and Latin and studied mechanics and mathematics. He continued his studies in experimental science at Oxford University, where he showed great talent for working on mechanical instruments. Such talent led him to become the assistant to one of the founders of modern chemistry, Robert Boyle. Boyle is most famous for his discovery of Boyle's law (the pressure exerted by a gas held at a constant temperature varies inversely with the volume of the gas). He made this discovery by a series of experiments using air pumps, most of which had been designed and built by Hooke. Hooke eventually became a professor at Gresham College, but even before that, his work with Boyle had made him quite well known. At age 27 he was appointed curator of experiments for the newly formed Royal Society, a position he would hold for the next 40 years.

While curator, Hooke made contributions in a wide variety of disciplines. He significantly improved the pendulum clock and invented the balance spring, which was critical to the accuracy of pocket watches (Christiaan Huygens independently invented the balance spring more than a decade later, which led to some disputes over credit). In 1660 Hooke discovered what became known as Hooke's law: the tension force in a spring increases in direct proportion to the length it is stretched to. He also performed many different and often cruel experiments on animals, alas something all too common at the time. This included designing an apparatus such that he could observe the beating of the heart and the inflating of the lung of a living dog. He also subjected himself to dangerous and often disgusting experiments. He medicated himself with various botanical purgatives, emetics, steel filings, mercury, absinthe, and foul mineral water. For a cousin of the King at one of the Royal Society meetings, he demonstrated the effects of the vacuum pump, which removed air from an enclosed chamber. Later, writing in the third person, he described what happened. "A man thrusting his arm upon exhaustion of the air had his flesh immediately swelled, so as the blood was neere breaking the vains, & unsufferable." Some of the Fellows examined the subject's arm afterwards and found it "speckled."

In a 1670 lecture, Hooke claimed that gravity applied to all celestial bodies and that the force of gravity between bodies decreases with the distance between them. If the force were to be removed, the celestial bodies would move in straight lines. This led to a priority dispute with Newton, who later discovered his universal law of gravitation. Newton's law states that two bodies in the universe attract each other with a force that is directly proportional to the product of their masses and inversely proportional to the square of the distance between them. Hooke claimed that it was his idea that had inspired Newton. But most scholars agree that it would take Newton's superior mathematical skills to actually prove the exact mathematical relationship. The two men also disagreed on the nature of light.

While curator, Hooke had essentially another career as an architect, working with his good friend and cofounder of the Royal Society, Christopher Wren. Hooke was appointed Surveyor to the City of London and made much more money as an architect than as a scientist, in part because he designed many of the buildings that replaced those destroyed by the Great Fire of London in 1666. Hooke was always a prickly character and became increasingly grumpy in his old age. Newton was not the only scientist he engaged in various feuds. His position as curator made him in some ways a servant to other members of the Royal Society, having to take their orders, something that he resented. Unlike Leeuwenhoek, he detested Oldenberg, the secretary of the Royal Society. Although a founding and important member of the Royal Society, by the end of his life Hooke had alienated many members. It is thought that several portraits were made of Hooke, but none exist. While Hooke was alive and a resident of Gresham College, the Royal Society had been able to use its facilities. However, upon his death the Society was told it had to relocate. Newton was now president, and one of the items that disappeared during the move was a portrait of Hooke. Historians do not think that anyone deliberately destroyed portraits of Hooke, but at the same time it does not appear that anyone made sure that they were saved.

Although Hooke's interests spanned many different disciplines, he is best remembered for his microscopic studies, which he published in a magnificently illustrated volume, *Micrographia*, in 1665. This book was a showcase for his particular talents: his skill as an artist, his understanding of both nature and light, and his outstanding abilities in designing and building scientific instruments. Although microscopes had already been around for some decades, Hooke improved on the single-lens microscope by building a compound one, containing a new screw-operating mechanism to bring the object into focus. Before that, the specimen had to be moved. In addition, he placed a water lens beside the microscope that focused the light from an oil lamp on to the specimen, greatly increasing how brightly it was illuminated. Hooke actually took over this work from Wren, whom he credited. Many of the drawings were, if not Wren's, certainly inspired by them. Like Leeuwenhoek, Hooke looked at everything: fleas, fish scales, eyes of a fly, razor blades, snowflakes, bodily fluids, and fossils. He explained the similarities and differences between petrified and modern wood with a description of the petrifaction process that is still respected by scientists today. Perhaps the first person to examine fossils under the microscope, Hooke suggested that fossils were actually the petrified remains of once living organisms. The original meaning of fossil was simply "dug up," and they were described in the context of mineral ores, natural crystals, and useful rocks. According to most seventeenth-century natural philosophers, fossils that resembled particular objects or organisms were made by "plastick virtue," a creative force within the earth capable of fashioning any shape out of stone. However, Hooke rejected the plastick virtue hypothesis. He identified microscopic "snail" shells. He saw how much the modern nautilus and the extinct ammonite resembled each other, and he concluded that the ammonite fossils had also possessed protective shells. However, he also noted differences between the smooth-shelled nautiluses and the corrugated-shelled ammonites, which caused him to ask: Where were the ammonites now? He suggested that just as Roman coins or urns are clues to interpret a past civilization, fossils are the "medals, urns or monuments of nature."[3] Yet Hooke's views on the origin of fossils were not fully accepted until the nineteenth century.[4]

Of all of Hooke's many descriptions, the one that has had the most lasting impact is what he observed viewing thin slices of cork under his microscope. He called the empty spaces between the wall-like structure pores or *cells*, and the name stuck. Thus Hooke is credited with discovering the building blocks of life. Yet the true significance of his findings would not be recognized for another 200 years. Throughout the 1700s, more and more plants were discovered to contain cells, but no one suggested that they were actually the basic constituents of the plant. The emphasis was on the cell wall that provided the structural component of the plant. The animalcules of Leeuwenhoek and the cells of plants were thought to be totally unrelated. Furthermore, there was no adequate explanation of where cells came from or how they were formed. There was little interest in the functional significance of cells. Most anatomists and physiologists were satisfied with describing tissues as being nervous, muscular, connective, and cellular.

---

3 Bowen, "The Scientific Revolution of the 17th Century," 14.
4 Rudwick, *The Meaning of Fossils.*

Micrographia is one of the most important scientific books ever written. Hooke used the book to expound on a variety of subjects, and it represented a systematic investigation of the vegetable, mineral, and animal kingdoms. Most importantly, the book showed what the microscope could do for understanding the biological world. With its absolutely beautiful, detailed woodcuts and descriptions of the microscopic world, it uncovered a part of the cosmos that no one had previously imagined. Just as with Leeuwenhoek, many people did not believe his drawings – the microcosm he showed was just too strange. But Hooke's work proves the old adage "Truth is stranger than fiction."

One would have expected to see a stampede to improve the quality of microscopes to reveal ever finer details of this fantastic world. Yet just the opposite happened. Instead, the microscope began to decline in importance in the late seventeenth century. It was difficult to use and still gave relatively poor images. One of the reasons the observations were disputed was that, when peering through the scopes, many other people couldn't see the same things. Furthermore, the animal microscopist's investigations were hampered by the fact that the tissues were soft and subject to rapid decay, and the constituents were usually of low contrast. The methods for preserving and preparing specimens were still quite inadequate. Hooke wrote in 1692 that very few people were still grinding lenses and he was right. Leeuwenhoek toiled on, but in increasing isolation. What eventually would become the most distinctive instrument for biologists became instead for a time a toy for the aristocracy and the subject of satire, along with the men of science who used it. In 1676 Thomas Shadwell's play *The Virtuoso* opened. The main character, Sir Nicholas Gimcrack, spent £2,000 on microscopes to learn about "the nature of eels in vinegar." Gimcrack also transfused sheep's blood into a madman who then bleated like a lamb, observed military campaigns on the moon, and read his Bible by the light of a rotting leg of pork. The play was poking fun at all of London's natural philosophers, but there was no doubt that Gimcrack was based on Hooke. Seeing the play for himself, Hooke was furious, writing in his diary "Damned Dogs.... People almost pointed."[5] The importance of understanding the detailed structure of a louse, the eye of a bee, or the

---

5 Quoted in Scott, "Robert Hooke."

structure of a plant was not understood by the general public, nor by most men of science as well.

With Hooke's many interests, it shouldn't be too much of a surprise that a book devoted to illuminating the microscopic world might also contain his theory of light. As an adherent to the mechanical philosophy, he regarded light as mechanical as well – pulses of motion through a material medium. He did not examine his theory in any depth, but some of his observations of colors through very thin films of various substances such as soap bubbles and mica meant that his ideas did play an important role in the history of optics. Having an accurate understanding of the nature of light would be critical to further improvements of the microscope. The microscope may have fallen out of favor temporarily, but in the nineteenth century it emerged as the most important tool for the biologist. It played a crucial role in understanding the true significance of Hooke's "cell." Indeed, advancements in microscopy raised new questions and created new controversies, and in doing so it must be considered one of the leading actors in the ongoing story of cell theory and the role it played in understanding life.

CHAPTER TWO

The Physical Basis of Life

There is some one kind of matter which is common to all living beings, and that their endless diversities are bound together by a physical, as well as an ideal, unity.

Thomas Huxley, "On the Physical Basis of Life," 1869

CELLULAR IDEAS BEFORE CELL THEORY

What makes something alive? How did the discovery of the cell shape investigations to unravel this mysterious quality we call life? Did life arise with the first cells, or is life older than cells? Although interest in microscopic studies declined for a period, the structures revealed by the microscope, such as Hooke's cell, raised important questions regarding the organization of living tissues. Hooke had observed similar structures to his cork cell in other plants and became convinced that they had functional significance, but exactly what that function was remained unclear. He thought they might be channels to carry fluids through the body of the plant, the way arteries and veins do in animal forms. Italian microscopist Marcellus Malphigi and the Englishman Nehemiah Grew also turned their microscopes to plants and found cell-like structures, calling their work "vegetable anatomy." In *Anatomes plantarum*, Malphigi mentioned "**utricles**" (or cells) and the "basic utrical structure" of plants. Grew investigated the question that Hooke raised, whether plants had circulatory systems similar to that of animals. Like Hooke, he was also unable to find analogous vessels to the arteries and veins or the valves that Harvey had identified in animal veins. After presenting his detailed descriptions of cells to the Royal Society, Grew later wrote them up in *The Anatomy of Plants.* His book and Malphigi's treatise extended our knowledge of the microanatomy of plants with their wonderful detailed drawings, but it also meant that the cell was discussed primarily as a structural unit, its functional significance still obscure. It was embryologists who, in their quest to understand development, paved the way for understanding the functional importance of cells.[1]

The microscope was critical to the investigation of living matter, but its role also illustrates an important debate that has underlaid the historical treatment of many scientific episodes. Some historians argue that it is advances in technology that determine and shape our understanding of nature. Others maintain that ideas are prior and drive the creation of specific technologies to investigate particular questions. Parsing it in this way is useful to clarify what issues are relevant in providing a complete account of the subject, but it is less useful in providing a more nuanced account if the story being told is from only one of these perspectives. Advances in microscopy have played an absolutely essential role in our understanding of the cell, and continue to do so today. At the same time, what we "see" always has to be interpreted. Maura Flannery has documented that the construction of cells is an artistic process. Stains and different kinds of microscopes with different wavelengths of light make possible the visualization of certain things but not others.[2] The history of cell theory shows that results were interpreted through not just the lens of a microscope, but the lens of particular ideas and prior philosophical commitments. Hooke, Grew, and Malphigi may have found cells in a variety of different plants, but other investigators asked what exactly were cells. Were they the basic constituents of plants, or were they part of some interconnected fabric? Were they common or rare? Did animal tissue contain cells? Leeuwenhoek undoubtedly observed red blood cells, but he did not make a connection that they

---

1 Gilbert, "Embryological Origins," 307–51.
2 Flannery, "Images of the Cell," 195–204.

were homologous to the structures he observed in his investigations of seeds and stems of plants. He thought that animal tissues were made up of "**globules**," but he did not think they had any resemblance to his animalcules. That was a topic that would be debated for some time. As noted above, many people had trouble even "seeing" what Leeuwenhoek and Hooke had observed. They suggested that these "cells" were nothing more than artifacts produced by the microscope. Nevertheless, the assumption that the material world was mechanistic and, therefore, knowable drove the development of better instrumentation. Whatever the ideas of different investigators, they all shared an underlying belief that improvements in the magnifying and resolving power of the microscope would reveal Nature's secrets. By the end of the 1700s, however, some investigators suggested that the cell might not even be the most basic structural component, much less the smallest independent entity imbued with that mysterious quality of "life" – an idea that is beginning to gain more and more traction in modern-day research.[3]

In the search for the fundamental constituents of living tissue, two ideas that gained currency in the early eighteenth century were the aforementioned "globules" along with "fibrilles," which had the property of irritability. While the idea of irritable fibers had been put forth earlier, the Swiss animal physiologist Albrecht von Haller (1708–77) was by far the most influential in claiming that the basic units that made up the body were fibers: "For the fiber is for the physiologist what the straight line is for the geometrician and from this fiber all shapes surely arise."[4] He thought that the larger fibers he observed were made up of smaller ones, which were linear arrays of atoms that were held together by gluten and shaped by pressure. These microfibers consisted of three types, according to Haller. First was the *tela cellulos* (cellular tissues), which made up the framework that supports the body. Second was the *fiber muscularis*, which had the property of "irritability." *Fibra nervosa* had the property of sensibility. With these three kinds of fibers, Haller explained all the various functions of the different animal organ systems, although he said little about the microscopic anatomy of either epithelial or glandular tissues. Many eighteenth-century

---

3 See chapter 7.
4 Haller, in Harris, *The Birth of the Cell*, 18. For original, see Haller, *Elementa physiologiae corporis humani*.

natural philosophers accepted this classification, or at least the idea that the fibers were the fundamental constituents of animal organs. The cellular structure observed in plants was regarded to be entirely different. Marie François Xavier Bichat (1717–1802) also thought that fibers were the ultimate constituents of animal tissue. In his investigations he identified 21 different kinds of fibers that formed two different types: simple ones and ones that were combined and interwoven. Although he is considered the father of **histology** (the microscopic study of cells and tissues), ironically he was untrusting of the microscope and thought that much of what was observed were merely optical artifacts. All of his observations were based on what he could observe by careful dissection, maceration, and simply using a hand lens. Thus, in the 1700s there was still wide disagreement on what was the smallest building block of life. The "fiber" theory as put forth by Haller and Bichat was generally accepted by the late 1700s for animals, although not for plants. Instead, vessels often had been observed in plants. And the microscope, although it had been around since Leeuwenhoek, still wasn't entirely trusted as reliable either.

Someone who attempted to draw parallels between animal and plant tissues was Kaspar Friedrich Wolff (1733–94). His ideas were heavily influenced by Grew, as he thought the fundamental unit was a globule or vesicle, which he also sometimes referred to as a cell. The fibers and vessels were secondary structures. Like Grew, he thought the vesicles were spaces formed in the growing tissue and initially contained air. However, sometimes the vessels became filled with sap, while the fat globules that he observed in animal tissues stored excess nutrients. In plants he had noticed smaller vesicles between the larger ones. He concluded that the growth of the plant occurred by the production of new vesicles and also by the expansion of old vessels. He considered the fibers and vessels to be secondary structures because they were not needed to produce new vesicles. Describing the growth of the seed leaf he wrote, "The young leaf arising from the seed turns out to be composed entirely of vesicles and is patently quite devoid of fibers, vessels or grooves of any kind. The new vesicles simply arose *de novo* within a glassy ground-substance.... They arise from a hitherto unadulterated, homogenous, glassy substance, without any trace of vesicles or vessels." Animal tissues were formed in essentially the same way: "The constituent elements of which all parts of the animal body are at first composed are globules which can always be made out with a modest microscope."[5] Yet some have claimed that Wolff may never have observed animal cells due to the inadequate resolving power of his microscopes.[6] Wolff's globule theory was regarded as in opposition to the fiber theory of Haller, whose considerable prestige meant that Wolff's work was unappreciated in his own time.

One other theory that attracted a lot of attention at the time was put forth by the Italian Felice Fontana (1720–1803). The first part of his major treatise was published in Italian in 1765, but a more complete French version was published in 1781, and an English translation of the French one appeared in 1787. The *Treatise on the Venom of the Viper* was primarily about snake venom and other poisons, but it included a section on the structure of animal tissues. Fontana claimed that with the possible exception of certain membranes, all animal tissue was composed of twisted cylinders. He observed these structures in many different types of animal tissue. He was convinced that these twisted cylinders were the fundamental subunit of animal tissues, writing,

> The primitive twisted cylinders that I found in the cellular tissue of nerves, tendons, and muscles are to be found in all the parts and organs that I know of. They are much smaller than the smallest of the red vessels which let through only one blood corpuscle at a time. All attempts I have made to break them down into cylinders of smaller size have failed.[7]

However, even if Fontana couldn't actually observe smaller cylinders, he went on to claim, "I know of no part of the animal body that doesn't show twisted cylinders if it contains cellular tissue." In this context cellular tissue probably meant connective tissue, and he may have even seen connective tissue fibrils. However, what is significant is that he also claimed that all tissue must be made up of these twisted cylinders, in spite of his own drawings clearly showing globules from a preparation he made from the skin of an eel! To a modern eye some look like flattened epithelia cells showing nuclei. However, if the red cell was a

---

5 Wolff, in Harris, *The Birth of the Cell*, 20–21. For original see Wolff, *Theoria generationis*.
6 Harris, *The Birth of the Cell*, 21. See Studnicka, *Acta. Soc. Scient. Natural Moravicae*, fasc: 4: 1.
7 Fontana, in Harris, *The Birth of the Cell*, 21–22.

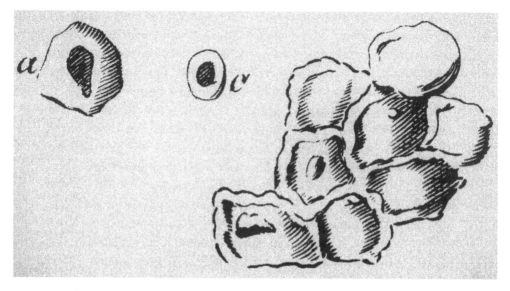

2.1 Fontana's globules from the skin of an eel, a single cell at higher magnification (a) with a red cell (c) for comparison.

human blood cell, the central body could not have been a nucleus and was probably an artifact of the preparation (see figure 2.1). Although widely discussed at the time, Fontana's theories were not widely accepted and later researchers did not appear to follow up on them.

All of these various ideas illustrate that the arrow between observation and theory points both ways. Observation gives rise to theory, but observations were theory-laden as well. Wolff's primary interest was to show that there was no preformation of the embryo. In trying to understand how organisms developed, he provided us with detailed observations of the structures of both plants and animals. In many ways Wolff's ideas can be considered a forerunner to the cell theory of Schleiden and Schwann, but we have no documentation that his research directly influenced them. Yet, as will be discussed further, Thomas Huxley gave priority to Wolff and was also critical of the cell theory of Schleiden and Schwann. Cells may have been observed in plants and animals, but no one thought that cells were the fundamental constituent of organisms. Hooke is credited with discovering the cell, but it would be two centuries before the importance of the cell was truly realized. Plants and animals shared the quality of being alive, clearly different from a rock, but, again, no one thought to emphasize the importance of the protoplasm in both animal and plant cells. In 1852 the Polish-German embryologist Robert Remak (1815–65) used the word "protoplasm" for the egg yolk and argued that it was the substance of all embryonic animal cells. Previously the term had only been used for plant cells. In 1855 the Austrian Franz Unger (1800–70) also remarked on the similarity between the sarcode of the zoologists and the protoplasm of the botanists. And in 1857, the German zoologist and comparative anatomist Franz Leydig (1821–1908) maintained that the cell could be a complete morphological unit, even in the absence of a cell wall. Thus, many researchers thought that this protoplasm might be the "stuff" of life. However, it was the German cytologist Max Schultze (1825–74) who developed the idea into a full-fledged theory – the **protoplasmic theory** of life.

THE PROTOPLASMIC THEORY OF LIFE

Cells had been found in both animal and plant tissues, consisting of a nucleus, protoplasm, and in plants a cell wall. However, this wall was not observed in animal cells, which was not surprising, since animal cells often moved and a rigid wall would impede such movement. Nevertheless, in some cases a cell membrane had been observed, but if one thought of a membrane as somewhat rigid, it would also inhibit such movement. In addition, creatures such as slime molds were multinucleate and no membrane was observed. Schultze thought that the membrane was actually an artifact that only became visible as result of the hardening of the protoplasm in the preparation of the cell for microscopic examination. Furthermore, based on his observations he claimed that cells undergoing division did not have a membrane. Even if a membrane existed and separated the contents of the cell from the external environment, he maintained that it was not really necessary since the unique protoplasmic substance did not mix with water. His observations of the movements of many diatoms convinced him that the original cell was membraneless. For Schultze the crucial aspect of the cell was the protoplasm, with the nucleus playing a still unknown role.

Given the contractibility of the protoplasm alterations in the shape of the whole cell are naturally impeded, if not

that the source of that vitality was in the cell. In fact, the prevailing theory remained that the origin of this vitality would never be found in the structure of the organism, no matter how powerful one's microscope. Rather, all living tissues were imbued with some immaterial vital force.

One of the most important characteristics of something considered to be alive was that it was capable of reproducing, and for higher organisms that involved mating. In a series of letters to the Royal Society over a period of about 20 years, Leeuwenhoek described his animalcules "coupling" as well as giving birth. In rainwater he observed minuscule creatures and "saw them copulate, the larger dragging the smaller through water after them, swimming by the means of very small fins." In another letter he wrote, "I observed some which were very much bigger than the rest and were coupled together, in which action they lay very still on the sides of the glass." Continuing his description, "another *Animalculum* that had brought forth two young ones, had her body laden with another sort of little creatures."[8] What started as a possible interpretation of his observations became "fact" for Leeuwenhoek over the years. He was convinced that his little animals copulated and that there were both males and females. While some researchers such as John Needham (1731–81) accepted his views, others such as Charles Bonnet (1720–93) were skeptical, and still others such as Lazzaro Spallanzani (1729–99) simply did not believe it. What Leeuwenhoek probably observed was not copulation, but binary fission. However, Abraham Trembley (1710–84) is credited with being the first to describe binary fission in an organism (see figure 2.2). Once again, the Royal Society played a critical role in promulgating various ideas about the nature of life. In a letter dated 6 November 1744 to the president of the society, Trembley described in detail the division of several different species of newly discovered microscopic freshwater polyps. They swam about, but eventually fixed themselves to solid surfaces. Prior to division the polyp withdrew its protuberances, rounding up into a ball, and then it "gradually splits down through the Middles, that is the middle of the head to the place where the posterior end joins the pedicle."[9] Finally, the newly formed "polypi" opened up, showing their "lips" and other distinguishing features. At first they were smaller than the polyp from

8 Leeuwenhoek, in Harris, *The Birth of the Cell*, 56.
9 Trembley, in Harris, *The Birth of the Cell*, 56–57.

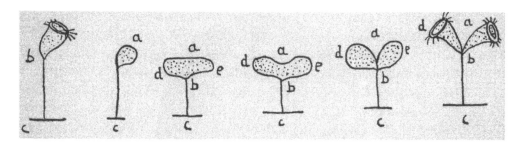

2.2. Trembley's illustration of binary fission in *Synedra*. Reproduced from John Randal Baker, *Abraham Trembley of Geneva: Scientist and Philosopher, 1710–1784* (London: Edward Arnold, 1952).

which they came, but within about an hour they grew to the same size. Over the next few years Trembley continued to observe this fission in several other colonial species. Yet it would still be another 100 years before it was recognized that binary fission was the general mechanism of reproduction for all cells, including multicelled animals and plants.

We see that until the 1800s microscopic investigations of the "cell" were primarily driven by structural considerations. Although Leeuwenhoek and Trembley observed their little organisms dividing, they did not think that they had any connection to the cells of Hooke or Grew, or to the fibrilles of Haller and Bichet. Wolff thought embryos were made up of globules, but his globules did not really correspond to what had been described by Hooke and others as cells or to some other basic structural unit. It was not until 1839 that Schleiden and Schwann formally articulated what has come down to us as cell theory. They claimed that all organisms, both plants and animals, were made up of cells, and that, in addition, the cell was the smallest independent basic unit of life. Yet clearly many people had been observing and describing cells far before 1839. Furthermore, by the end of the eighteenth century several researchers started to move their investigations into what was *inside* the cell. This began what will be a critical shift in analyzing the cell in terms of both structure *and* function. The French cytologist Félix Dujardin (1801–60) investigated various **infusoria**, that is, single-celled organisms found in infusions of decaying organic matter. In 1835 he described the material inside the microscopic organisms as a glutinous diaphanous substance that was insoluble in water and contracted into a spherical mass, sticking itself to the dissecting needles, that could be drawn out

like slime. He called this substance a **sarcode**. In 1839 the Bohemian (Bohemia later became part of the Czech Republic) researcher Jan Purkyně (1787–1869) confirmed what Dujardin found, but he called the thick fluid inside the cell wall **protoplasm**, and that is the name that became accepted. Purkyně made many important discoveries, once again due in part to advancements in microscopy. Although Wilhelm His is often credited with the invention of the **microtome**, Purkyně is usually credited as being the first person to use it extensively in his research. The microtome allowed him to make very thin slices of tissue, which he then examined with a new, improved compound microscope. Purkyně is perhaps best known for his 1837 discovery of large neurons with many branching dendrites in the cerebellum that now bear the name "Purkyně cells." Two years later he discovered Purkyně fibers, the fibrous tissue that conducts electrical impulses from the atrioventricular node to the ventricles of the heart. These important discoveries, as well as many others concerning animal tissues, overshadowed Purkyně's contributions to cell theory, which as a result have been underappreciated. However, Purkyně had begun his experimental work by studying the microstructure of plants. He was fascinated with the dynamic nature elastic fibers that caused the anther to burst and release its pollen. along Purkyně was interested in the structure and function of both and animal cells, and he often commented on the similarity between both kinds of cells in his writings.

The German botanist Hugo von Mohl (1805–72) provided evidence of the importance of the viscous fluid inside the cell a us our modern definition of protoplasm. However, the Germ nist Ferdinand Cohn (1828–92) was the first person to expli in 1850 that the protoplasm of plants and animals was virt tical. The motile stage of a unicellular green alga with its toplasmic cilia reminded him of ciliated unicellular anim *Euglena*. The contractile substance found in the sarcode a plant cell's ability to make vacuoles led him to conclude *toplasm* of the botanists and the contractile substance or zoologists, if not identical, are at all events in the highes gous formations."[10] Throughout the 1850s a series of p

---

10 Cohn, in Geison, "The Protoplasmic Theory of Life," 275. For "Zur Naturgeschichte des Protococcus pluvialis," 605–764.

altogether rendered impossible, by the presence of a rigid cell membrane. The less the surface of the protoplasm is hardened into a membrane, the nearer the cell approaches its original membraneless state in which it is no more than a naked blob of protoplasm containing a nucleus.[11]

Because both plant and animal cells shared this protoplasm, he thought it was the basic unit of life rather than the cell itself. The German physiologist Ernst Wilhelm von Brücke (1819–92) agreed and wrote in 1861 that the cell wall was not necessary; it was the contents inside the cell that were responsible for life. He thought that developments in histology would reveal that the cell concept of Schleiden and Schwann would need to be abandoned. Throughout the 1860s several biologists, including Ernst Haeckel, Heinrich Anton de Bary, and Wilhelm Kühne, working on a variety of different organisms, published important papers emphasizing that protoplasm was the key component to defining life.

While little if any mention is made of the protoplasmic theory in today's biology textbooks, historically these ideas played an absolutely critical role in furthering our understanding of "life." As Gerald Geison wrote, "The publication of [Schultze's] paper, more than any other single event, marked the birth of the protoplasmic theory of life. On physiological rather than structural grounds, and with special emphasis on the properties of contractility and irritability, Schultze demonstrated that a single substance, called protoplasm, was the substratum of vital activity in the tissues of all living organisms, however simple or complex."[12] It was Thomas Huxley, however, who brought widespread attention to Schultze's ideas, exposing them to a much larger audience. Huxley gave himself the name "Darwin's bulldog," and it stuck. Many people only know him in that capacity. But for our purposes, we should call him Schultze's bulldog,[13] which also reflects more accurately his own primary research interests and helps illuminate his criticisms of cell theory. In Edinburgh on 18 November 1868, Huxley inaugurated a Sunday evening lecture series on nonreligious topics with his lecture

---

11 Schultze, in Harris, *The Birth of the Cell*, 151–52. For original see Schultze, *Archiv für Naturgeschichte*, 287.

12 Geison, "The Protoplasmic Theory of Life," 276–77.

13 I thank Douglas Allchin for this suggestion.

"The Physical Basis of Life." By February of the next year, Huxley's lecture was the lead article in the London *Fortnightly Review*. It was widely reprinted over the next several years in a variety of different publications both in Britain and abroad. Huxley had already made a name for himself as the foremost advocate for Darwin's theory. His book *Man's Place in Nature* (1863), also based on a series of public lectures but greatly expanded, argued eloquently and powerfully that humans were no exception to Darwin's theory. Extremely successful, it had been reprinted several times by 1868. Interestingly, however, the enormously popular Victorian periodical, *Vanity Fair*, did not even mention Huxley's defense of Darwinism in their portrayal of him. *Vanity Fair*, "*A Weekly Show of Political, Social and Literary Wares*," was the leading society magazine of the time, best known both then and now for its caricatures of prominent personages with satirical commentary. From the time of its inception in 1868 until it ceased publication in 1914, more than 2,000 portraits, done by a group of international artists, appeared in its pages. They provide a rich pictorial history of the times. The Italian Carlo Pellegrini, who worked under the pseudonyms "Singe" or "Ape," drew Huxley's portrait in 1871, which was number 19 in their "Men of the Day" series. Titled "Professor Huxley," with the caption "A great Med'cine Man, among the Inqui-ring [*sic*] Redskins," it also claimed that he was the "inventor of protoplasm."[14]

As will be analyzed in greater depth, the ideas presented in "The Physical Basis of Life" provide insight into Huxley's objections to certain aspects of cell theory and will be one of several threads that connects the narrative of this book. In spite of the primacy that is given cell theory in virtually all standard histories of biology, a strong case can be made that in the nineteenth century, the "protoplasmic theory" rather than cell theory was what furthered the structural/functional debates over the role of the cell for understanding both development and heredity. And in doing so it set the agenda for modern cell biology. As later chapters will explore, the present-day interest in a more holistic, systems-level approach to understanding the structure and function of cells can trace its origin back to the concept of protoplasm.[15]

14　Blinderman and Joyce, The Huxley File.
15　Welch and Clegg, "From Protoplasmic Theory to Cellular Systems Biology," C1280.

Indeed, much current research on the role of cytoplasmic organization suggests that cell theory should be revised. As was often the case in Huxley's career, his objections to various theories represented a minority viewpoint, but he raised key issues that were problematic and are at last being addressed in modern-day research. How did Huxley come to be such an advocate of protoplasmic theory as well as so highly critical of cell theory?

THOMAS HUXLEY: REBEL WITH A CAUSE

Thomas Huxley was born in 1825, the youngest of seven children. His father was a schoolteacher, but Huxley received little regular schooling and was largely self-taught. Although he attended church regularly, as he grew older he realized that he was one of those skeptics or infidels that preachers spoke of with horror. Even as a young boy he had interests in a staggering array of subjects. In spite of often being called an atheist or a materialist, he vehemently denied such labels. Later, as a member of the Metaphysical Society, where every variety of philosophical and theological opinion was expressed and most of his colleagues were -ists of some kind, he described himself as a "man without a rag of a label" and claimed he had some of the same uneasy feelings that "beset the historical fox when after leaving the trap in which his tail remained he presented himself to his normally elognated companions." He coined the word "agnostic" to describe his own philosophical system of belief, being "anti-thetic" to the word, "gnostic" of Church history, who supposedly knew so much about the topics of which he was ignorant. He then "took the earliest opportunity of parading it at our Society, to show that I, too, had a tail like the other foxes."[16] Huxley showed an affinity for anatomy and entered medical school when he was only 15, receiving a scholarship to the medical school attached to Charing Cross Hospital. Except for physiology, most of the medical curriculum bored him. Nevertheless, he won first prize for chemistry, also taking honors in anatomy and physiology. He later won a gold medal in those subjects in the Bachelor of Medicine exam. When he published the first research paper on the root sheath of hair that is still known

---

16 Huxley, "Agnosticism," 239.

as Huxley's Layer, he was too young to qualify for a license to practice medicine!

Like many others who made their mark in the natural sciences, Huxley took a voyage around the world. He was the assistant surgeon on the HMS *Rattlesnake* from 1846 to 1850. The voyage also resulted in some of his most important scientific work and established his reputation within the scientific community. However, it was his defense of Darwinism that brought him into the public spotlight. At the 1860 meeting of the British Association for the Advancement of Science, when asked by Bishop Wilberforce whether it was on his grandfather's or grandmother's side that the ape ancestry comes in, Huxley famously replied

> that a man has no reason to be ashamed of having an ape for his grandfather. If there were an ancestor whom I should feel shame in recalling it would rather be a man – a man of restless and versatile intellect – who, not content with an equivocal success in his own sphere of activity, plunges into scientific questions with which he has no real acquaintance, only to obscure them by an aimless rhetoric, and distract the attention of his hearer from the real point at issue by eloquent digressions and skilled appeals to religious prejudice.[17]

Huxley's response created pandemonium. According to one account a lady fainted. While various versions exist of the encounter, they do not differ in substance. As his son Leonard Huxley wrote, it was an important milestone in his father's career. Not only did it ensure that Darwin's theory received a fair hearing, but it let the public know that Huxley was a force to be reckoned with in the world of science and religion.

More than anyone else, Huxley was responsible for disseminating Darwin's theory to the English-speaking world. However, Huxley had a research program established long before he went to battle against the enemies of evolution, and he continued to publish monographs in virtually every area of zoology. In spite of dubbing himself "Darwin's bulldog," he was skeptical of the two basic tenets of Darwin's theory: natural selection and gradualism. Pre-*Origin* Huxley did not believe

---

17 Rev. W.H. Freemantle, 1892, in Huxley, *Life and Letters of Thomas Henry Huxley*, 200. The different accounts are reprinted in this volume. See also Lyons, *Thomas Henry Huxley*.

in transmutation. The research of Karl Ernst von Baer and Georges Cuvier, as well as Huxley's own research, suggested that organisms could be grouped into discrete types and that no transitional organisms existed between them. While his early work from the *Rattlesnake* superficially appears to be merely a series of detailed monographs on various invertebrates, he had a much more ambitious agenda. He wanted to provide a theoretical foundation for taxonomy in order to understand how form came to be generated. It is this interest that brings him to cell theory, and why his critique of it is so important in the history of the cell. Crucial to building a taxonomy was the concept of type. With Darwin's publication of *On the Origin of Species* in 1859, Huxley recognized that the unity of type he was observing was due to descent from a common ancestor. As he wrote Darwin, "It is true that the more widely two animals differ from one another, the earlier does their embryonic resemblance cease, but you should remember that differentiation that takes place is the result not so much of new parts as of the modification of parts already existing and common to both of the divergent types."[18] Furthermore, as he later wrote, "Evolution is not a speculation, but a fact; and it takes place by epigenesis."[19] However, Huxley also maintained that the fossil record did not document gradual change. Rather, there were great gaps, and evolution appeared to occur in jumps or by saltation. Darwin agreed, writing that the absence of intermediate fossil forms "is the most obvious and gravest objection which can be urged against my theory."[20] He devoted two chapters in *Origin* to explaining that while the fossil record might not be used in support of gradualism, it should also not be used against it. Huxley's advocacy of the type concept also contributed to his advocacy of a saltational (from the Latin verb *salire*, meaning to leap or jump) view of evolution. Saltation allowed him to explain the gaps in the fossil record, accept Darwin's theory, and also maintain his belief in the unity of type.[21] As will be explored in later chapters, Huxley's caution about the fossil record received its most serious hearing through the theory of **punctuated equilibrium** of Niles Eldredge and Stephen Gould. They

---

18 Huxley, 7 July 1857, in Darwin, *The Correspondence of Charles Darwin, 1856–1857*, vol. 6, 426.
19 Huxley, "Evolution in Biology," 288.
20 Darwin, *On the Origin of Species*, 292.
21 Lyons, *Thomas Henry Huxley*.

argued that the gaps in the fossil record were real, not just the result of an imperfect record. Instead, the fossil record documented a pattern of periods of rapid diversification interspersed with periods of relatively little change or stasis. That in turn played a key role in the emergence of the discipline of evolutionary developmental biology, affectionately known as evo-devo. Evo-devo finally started to address the issues that Huxley raised by incorporating the findings from embryology into evolutionary theory. To fully understand how an egg becomes a chicken, a broader perspective than just natural selection is needed. Development also cannot simply be explained by focusing solely on genes. Likewise, biologists are now recognizing that Huxley's views on cell theory are worth revisiting and of interest to a historian of biology as well.

Huxley's advocacy of Darwin's theory served another purpose in his multifaceted agenda of advancing knowledge. He wanted a clear line drawn between science and religion. Darwin's theory provided a completely naturalistic explanation of the history of life, free from any argument of design or supernatural causation. Huxley had no use for vitalistic theories on the origin of life. As the caption of *Vanity Fair's* illustration of Huxley proclaimed, "He refuses to believe in angels, because the telescope has not yet discovered them. Like a man who hops on one leg, instead of walking erect with his face heavenwards, he has to pick his steps with care through the mud of Materialism, and in this way it has come to pass that he has stumbled on Protoplasm."[22] *Vanity Fair's* caricature of Huxley appeared three years after Huxley's lecture "On the Physical Basis of Life," which threw the religious community into an uproar.

THE PHYSICAL BASIS OF LIFE

Appearing before a large Edinburgh audience with a bottle of smelling salts, water, and various other common substances, Huxley claimed that he had all the basic ingredients of protoplasm, which he translated as "the physical basis of life." He told his audience that matter and life were inseparably connected. The lichen, the pine, the fig, a Finner whale, the flower in a girl's hair, and the girl all exhibited a threefold

---

22 *Vanity Fair*, 28 January 1871.

unity: "a unity of power, a unity of form, and a unity of substantial composition."[23] By unity of power, Huxley meant that all the activities of living organisms, from amoebas to humans, fell into three basic categories. They were involved in (1) maintenance and development, (2) the continuation of the species, and (3) effecting changes in the relative position of parts of the body. Huxley went even further, claiming:

> Even those manifestations of intellect, of feeling, and of will, which we rightly name the higher faculties, are not excluded from this classification.... Speech, gesture, and every other form of human action are, in the long run, resolvable into muscular contraction, and muscular contraction is but a transitory change in the relative positions of the parts of a muscle.[24]

All living organisms also shared a common unity of form, the cell – and all cells shared a similar chemical composition. Plants and animals appeared to be very different, yet no sharp dividing line existed between even the simplest of these organisms. Huxley went on to make an even more radical claim: the distinction between living and nonliving matter could be found in the arrangement of molecules.

> All vital action may ... be said to be the result of the molecular forces of the protoplasm, which displays it, and if so, it must be true, in the same sense and to the same extent, that the thoughts to which I am now giving utterance, and your thought regarding them, are the expression of molecular changes in the matter of life which is the source of our other vital phenomena.[25]

Based on these comments, one would think that Huxley would have embraced cell theory wholeheartedly. But that was not the case.

For Huxley, it was not the cell itself that was responsible for life, but rather the contents inside. He pointed out that the stinging property

---

23 Huxley, "On the Physical Basis of Life," 131–33.
24 Huxley, "On the Physical Basis of Life," 133–34.
25 Huxley, "On the Physical Basis of Life," 154.

of the common nettle was due to delicate hairs that covered its surface. The protoplasm of each hair was full of granules that were constantly in motion. Contraction of the whole thickness of the substance passed from point to point, giving rise to the appearance of progressive waves and resulting in a streaming movement of the granules. Currents could be observed going in both directions. In many ways the motion was very similar to the internal circulation of blood observed in animals. For instance, in certain circumstances the protoplasm of the alga or fungus became free of its woody case and the whole organism moved, propelled by the contractibility of one or more hair-like projections of the body. Although it had not been definitively proven, the phenomenon in animals and plants appeared to be similar. As Huxley pointed out, the simplest organisms had no nucleus. A single mass of protoplasm was the entire organism. Thus he asked, "How is one mass of non-nucleated protoplasm to be distinguished from another? Why call one a 'plant' and the other 'animal'?"[26] Huxley was not denying that there was an enormous difference between the lowest and the highest organisms or between plants and animals, but he argued that the difference was only "one of degree, not of kind." Life was an emergent quality that lay in the arrangement of the molecules. "Under whatever disguise it takes refuge, whether fungus or oak, worm or man, the living protoplasm not only ultimately dies and is resolved into its mineral and lifeless constituents, but is always dying, and strange as the paradox may sound, could not live unless it died."[27] Claude Bernard had essentially made the same claim: "*La vie, c'est la mort.*" The fundamental property of life was the ability of the organism to respond to external agents, and this was shared by both plants and animals. Huxley agreed, arguing there was no essential difference between animals and plants. He made a powerful case for the unity of life. All organisms were made up of cells. However, he emphasized the contents inside the cell – the protoplasm – as the source of vitality, rather than the cell itself. Furthermore, just as it was the arrangement of molecules inside the cell that gave the property of life, likewise it was the arrangement of the cells within the organism that gave the organism its properties. Yet somewhat paradoxically, it was this idea that resulted in him being

---

26 Huxley, "On the Physical Basis of Life," 141.
27 Huxley, "On the Physical Basis of Life," 145.

critical of cell theory. The cells were *not* anatomically independent. Rather, they were interconnected, creating larger aggregates and communicating with one another. Therefore, cells could not be the elementary units of life.

I have described Huxley as a morphologist because he was interested in the problem of form; for Huxley the answer to how a particular form came to be would be found in understanding development. Trying to unravel this complicated process of how a single cell gave rise to a multiplicity of different cell types with specialized functions resulted in one of the most fundamental debates in nineteenth-century biology: did structure determine function or the other way around? For Huxley, the two ideas were inextricably linked. Nevertheless, in reality it was questions of physiology that most excited him. As he wrote in his autobiography, what "interested me was physiology, which is the mechanical engineering of living machines, ... what I cared for was the architectural and engineering part of the business, the working out of the wonderful unity of plan in the thousands and thousands of diverse living constructions, and the modifications of similar apparatuses to serve diverse ends."[28]

In 1853 Huxley wrote a detailed and penetrating review of Schleiden and Schwann's cell theory in which he explained why certain tenets of the theory would lead researchers astray in their further investigations. He began by giving a brief overview of the history of investigations concerning the cell. Underlying Huxley's review of the cell was a broader question: what was the most fundamental goal of biologic investigations? For Huxley it was to understand what constituted life. He suggested that the drive to understand the structure of living matter was undoubtedly motivated by the activity that was intrinsic to something that was "alive." It was what distinguished a rock from a fern. For the last 300 years, Huxley maintained, research into anatomy, physiology, and development had been more or less equally divided. He claimed that making real progress in understanding the physiology of an organism presupposed an understanding of structure. However, to really understand structure, one had to understand development. Huxley was particularly appreciative of Wolff's recognition of the importance of development in understanding both

---

28 Huxley, "Autobiography," 7.

structure and function. Many researchers had already realized that organisms were composed of comparatively few elementary parts, and that each of these individual parts had a certain vitality. That vitality was dependent on certain general conditions, such as proper nutrients being available, but each part was independent of the direct influence of other parts. Huxley claimed that these essential ideas of cell theory had been expressed three centuries earlier. Nevertheless, improvements in microscopy gave validity to ideas that previously had been highly speculative in many ways, especially in illuminating the nature of those individual parts as cells. For Huxley, the cell theory of Schleiden and Schwann was important for two reasons. First, when the research "by innumerable guideless investigators, called into existence by the tempting facilities offered by the improvement of microscopes, threatened to swamp science in minutiae and to render the noble calling of the physiologist identical with that of the 'putter-up' of preparations, the cell theory grouped together masses of detailed observations in a clear and organized manner."[29] Second, theory and hypothesis were essential to the process of science. Indeed, science could not progress without the continual production of new theories to guide experiments to get closer to the "the truth." He explained, "There are periods in the history of every science when a false hypothesis is not only better than none at all, but is a necessary forerunner of, and preparation for, the true one."[30]

While initially extolling the contributions of Schleiden and Schwann, in actuality Huxley's review was highly critical. As was typical of Huxley, a master of rhetoric, he even used Schwann's words against him. Schwann had claimed that his and Schleiden's work was different from previous researchers in that "[t]he theory of the present investigation was, therefore, to show ... that there exists a common principle of development for all the elementary parts of the organism."[31] However, Huxley gave priority to Wolff in realizing the importance of development, and there is much to be said for Huxley being correct in that assessment. Schwann had also claimed that "[a]n hypothesis is never hurtful, so long as one bears in mind the amount

29 Huxley, "The Cell Theory," 250.
30 Huxley, "The Cell Theory," 249.
31 Huxley, "The Cell Theory," 249.

of its probability, and the grounds upon which it is formed ... even though there be a risk of upsetting this explanation by further investigation; for it is only in this way that one can rationally be led to new discoveries, which may either confirm or refute it."[32] And that was exactly what Huxley set out to do, arguing that much of Schleiden's and Schwann's claims were "based upon errors in anatomy, and lead to errors in physiology."[33]

Huxley claimed that investigations by himself and others refuted several key components of cell theory:

1 The prevalent notion of the anatomical independence of the vegetable cell considered as a separate entity
2 The prevalent conception of the structure of the vegetable cell
3 The doctrine of the mode of its development

First, the individuality of each cell, according to Huxley, was an artifact due to the plant tissues being treated with strong reagents, along with mechanical processes resulting in them being "broken up into vesicles corresponding with the cavities which previously existed in it." In addition, based on the research of von Mohl, Huxley claimed, "there exists no real line of demarcation between one cell and another." If this was the case, cells were not anatomically independent, and that had important physiological implications. For Huxley, cells were interrelated and conveyed information, which meant that the cell was not the elementary unit of life. The quality of "life" or vitality did not lie in the structure of the individual cells, but rather in the processes themselves. To really understand life one had to look at development. According to Huxley, Schwann's organism was like a beehive with all of its parts (cells) independent but working together to create the various processes and forces that resulted in development. Huxley contrasted this with Wolff's view in which the organism was a mosaic. The primary elements were neither anatomically nor physiologically independent. They came about as a result of molecular forces that were a product of the whole organism, which played an essential role in directing development. The organism resulted from the differentiation of a primarily

32 Theodor Schwann, quoted in Huxley, "The Cell Theory," 249–50.
33 Huxley, "The Cell Theory," 251.

homogeneous whole into these parts. This, Huxley claimed, was the logical conclusion of how development must occur according to the general principle of epigenesis.

Another person whom Huxley greatly admired was Karl Ernst von Baer. Quoting von Baer, Huxley writes, "The history of development, he says, 'is the history of a gradually increasing differentiation of that which was at first homogeneous.'" Huxley asked why shouldn't we extend this view to histology, since that is only **morphology** at a finer level. If one adopts this position, then the cell structure precedes the more specialized structure, but it was not the result of a "cell force," as Schwann claimed. Rather, the cellular structure of organisms was "simply a fact in their histological development." He acknowledged that the cell was a histologically necessary stage, but there was no more a necessary causal connection between it and what came after it than the equally mysterious morphological necessity for the chorda dorsalis (the rod-shaped cord of cells on the **dorsal** aspect of an embryo, defining the primitive axis of the body) that preceded the development of true vertebrae.

It was still unclear what the role of the nucleus was, but Schwann and Schleiden, and also Albert von Kölliker, thought the nucleus played an absolutely essential role in the life of the cell. Huxley claimed that Schwann, based primarily on the work of von Kölliker, thought the cell had powers that were not due to separate molecules. Von Kölliker had suggested that the existence of a peculiar molecular attraction proceeding from the nucleolus first, and subsequently from the nucleus, was the ultimate cause of cell division. Huxley acknowledged that the nucleus was important to cell division, but he had an entirely different view about its role. While proponents of cell theory described the cell as consisting of a nucleus, protoplasm, and cell wall or membrane, Huxley described the cell and its constituents in a different way, inventing the new terminology of "endoblast" and "periplast"; the details need not concern us. However, what is important is his description of the changes in the cell as development proceeded. In plant cells, the constituents of the cell wall, such as cellulose, as well as the substance between the cells became converted into other substances; this created cavities and the intercellular channels observed in plants. In animal cells, vacuoles formed with various deposits and also vascular canals. In addition *fibrillation* – a tendency to break up in certain definite lines rather than others – occurred. For Huxley, what was crucial to this process was that the cells were not separate; they were interconnected, eventually forming special structures such as tissues and organs. In his scheme, the nucleus and protoplasm were secondary. Indeed, they were "almost accidental modifications." Huxley thus rejected the Schleiden-Schwann model of the cell. Schwann claimed that cells were "machines" that were necessary for further development to take place. Instead, Huxley claimed "that they are no more the producers of the vital phenomena than the shells scattered in orderly lines along the seabeach are the instruments by which the gravitative force of the moon acts upon the oceans. Like these, the cells mark only where the vital tides have been, and how they have acted."[34]

While many people thought there was much to be said about Huxley's position, he definitely was in the minority. He certainly was mistaken about the importance of the nucleus, and his terminology never caught on. Nevertheless, he raised issues that reflected an epigenetic perspective that would be increasingly important in understanding development. Emphasizing physiological processes over structure and putting the source of vitality in the activity of individual molecules reflected a more holistic view of the organism and meant that the individual cell was less important. As he wrote, "The individuality of a living thing, then, or a single life, is a continuous development, and development is the continual differentiation, the constant cyclical change of that which was, at first, morphologically and chemically indifferent and homogeneous."[35] His position also reflected his view against vitalism. No special vitalistic force animated life nor were specific structures necessary. "Life" emerged from the properties and arrangement of molecules alone.

Many historians consider the most important legacy of protoplasmic theory to be its contribution to the demise of vitalistic theories. Yet, as Huxley correctly pointed out at the time, protoplasmic theory also implied a critique of cell theory, and he was not alone in thinking this.[36] Alluding to Hooke's cell, the plant physiologist Julius von Sachs claimed that "to call the protoplasmic unit a cell was about as

---

34 Huxley, "The Cell Theory," 278.
35 Huxley, "The Cell Theory," 268.
36 Liu, "The Cell and Protoplasm," 889–925.

appropriate as calling a live bee in a honeycomb a cell."[37] He thought
the term should be abandoned. As he wrote,

> the science which deals with living matter commenced with
> a controversial word originating as a mistake more than 200
> years ago, and then maintained up to the present day: it is the
> word Cell. It is well-known that the word Cell should be un-
> derstood only from a historical perspective as Robert Hooke
> referred to the inner structures of cork as being cellular be-
> cause they closely resembled hexagonal wax cells of hon-
> eycomb. Later, zoology also accepted this unfortunate term
> despite the word being even more controversial when ap-
> plied to animals.[38]

Like Huxley, Sachs thought that the cell should be regarded as merely
providing housing for the "stuff of life." Instead, Sachs described something he named the "**energide**" and argued it should be considered the
smallest independent unit of life. This consisted of the nucleus with
its surrounding sphere of protoplasmic influence. Several other biologists were also critical of cell theory. The medical researcher Edmond
Montgomery pointed out the multinucleated nature of muscle tissue
and the apparent fusion of neurons in the nervous system meant that
the organism was not just an aggregate of individual cells. Rather the
organism consisted of "an unbroken living substance, or single physiological unit."[39] The anatomist Carl Heitzmann also rejected the idea
that higher animals were merely cell aggregates, citing evidence that
supposedly distinct cells remained connected after cleavages by many
fine protoplasmic "bridges."[40] The zoologist Charles Otis Whitman
(1842–1910), whose work we will discuss more thoroughly in chapter
4, citing the work of Huxley, Sachs, and others, argued that growth and
differentiation were not dependent on the formation of new cells by
division. Furthermore, the cell concept was interfering with what was

---

37 Julius von Sachs, 1892, quoted in Welch and Clegg, "From Protoplasmic Theory to
    Cellular Systems Biology," C1280–90.
38 Julius von Sachs, 1892, quoted in Baluška, Volkmann, and Barlow, "Cell-Cell
    Channels," 5.
39 Edmund Montgomery, 1881, quoted in Reynolds, *The Third Lens*, 38.
40 Carl Heitzmann, 1883, quoted in Reynolds, *The Third Lens*, 38.

being seen. "We are so captured with the personality of the cell that we habitually draw a boundary-line around it, and question the testimony of our microscopes when we fail to find such an indication of isolation."[41] The embryologist Adam Sedgwick, like Huxley and Whitman, was also extremely critical of cell theory, particularly as it related to development. It was a "fetish" holding "men's minds in an iron bondage." It "blinds men's eyes to the most patent facts, [and] obstructs the way of real progress in the knowledge of structure."[42] Like Huxley, Montgomery, Heitzmann, and others, Sedgwick cited evidence of protoplasmic bridges in the development of animals as well as plants. Once again we see (!) how what one sees is interpreted through the lens of the observer. As Sedgwick wrote, the cell is "a kind of phantom, which takes different forms in different men's eyes." Cells were not as distinct and isolated as cell theory claimed. The American cytologist Edmund Beecher Wilson also agreed. In spite of titling his very influential 1896 textbook *The Cell in Development and Inheritance,* he began his first chapter by asking students of science to stop using the word "cell." "The term 'cell' is a biological misnomer.... for whatever the living cell is, it is not, as the word implies, a hollow chamber surrounded by solid walls. The term is merely an historical survival of a word casually employed by botanists of the seventeenth century to designate the cells of certain plant tissues which, when viewed in section, give somewhat the appearance of a honeycomb."[43] Furthermore, in a footnote he claimed that Sachs's energide concept was more appropriate in both morphological and physiological senses, and that "[i]t is to be regretted that this convenient and appropriate term has not come into general use."[44]

More than 40 years later, in an even more damning critique, specifically of Schleiden and Schwann, the American cytologist Edwin Conklin (1863–1952) argued that these two men's "scientific legacies were best respected by expunging their names from the annals of cell theory altogether."[45] As we will see in the next chapters, critics of different aspects of cell theory appeared throughout the twentieth century, but they have always represented a minority viewpoint. However,

41 Whitman, "The Inadequacy of the Cell-Theory in Development," 645.
42 Adam Sedgwick, 1895, quoted in Reynolds, *The Third Lens*, 40.
43 Wilson, *The Cell in Development and Inheritance*, 13.
44 Wilson, *The Cell in Development and Inheritance*, 13.
45 Edwin Conklin, quoted in Liu, "The Cell and Protoplasm," 890.

a tipping point may soon be reached. Daniel Mazia's concept of the "cell body" described essentially the same idea as Sachs's energide, but with much more evidence to back it up. František Baluška has extended Mazia's research, claiming that the cell body/energide is the smallest independent unit of life. For other reasons as well, Baluška argues that cell theory needs to be revised. Both men's work will be explored more thoroughly in chapter 7.

There was a third component, however, to what has come down to us as the modern cell theory: all cells come from preexisting cells. This was contrary to the idea of **cell-free formation** that Schleiden and Schwann advocated. As our story unfolds, this aspect may turn out to be the most important legacy of the cell theory from the nineteenth century. The acceptance of the third tenet was crucial in understanding life processes. It meant not only that there was a continuity of development between the fertilized egg and the organism it became, but also continuity from generation to generation. Spontaneous generation did *not* occur. This was certainly something that Huxley agreed with. Furthermore, this third aspect contributed significantly to the acceptance of cell theory, which in turn was absolutely essential to furthering our understanding about both heredity and development, and is the subject of the next chapter.

CHAPTER THREE

The Cell as the Unit of Heredity and Development

A cell has a history; its structure is inherited, it grows, divides, and, as in the embryo of higher animals, the products of division differentiate on complex lines.

Sir Frederick Gowland Hopkins, "Some Aspects of Biochemistry," 1932

CELL THEORY

As the previous chapter showed, many people had been observing cells prior to Schleiden and Schwann. Wolff's globules were clearly cells. Some had even adopted Hooke's usage of the word "cell" in describing the structures they were observing in a variety of different tissues and organisms. In particular, Purkyně had on many occasions pointed out the similarity between animal and plant cells. It is apparent that he was not just referring to structural similarities, but to functional ones as well. Others were also bringing attention to what was inside of the cell and that those substances and processes were essential to understanding life. Furthermore, Schleiden and Schwann were wrong about how cells originated. It was, however, the third tenet – that cells only arise from preexisting cells – that will be critical to establishing the widespread acceptance of cell theory. As is so often the case in science, many initial theories are eventually shown to be wrong, but they still provide a useful framework for further investigations. In a further irony of history, as mentioned in the introduction and will be discussed below, Virchow is credited with the third tenet, but Robert Remak had demonstrated several years earlier that cells only come from preexisting cells. So why are Schleiden and Schwann considered the originators of cell theory? Their theory of cell-free formation was eventually shown to be incorrect. Nevertheless, it brought attention to the nucleus. Purkyně may have claimed that plant and animal cells were similar, but he did not discuss how cells were generated and he did not pay much attention to the nucleus. Emphasizing the importance of the nucleus was also critical to acceptance of cell theory.

This chapter explores how the ideas of Schleiden and Schwann furthered investigations of the nucleus, bringing attention to the role it played in understanding both heredity and development. It again shows how the prestige of particular scientists influences the acceptance or rejection of particular ideas. In many ways, it might have been better to refer to it as the nuclear theory rather than the cell theory, as research became increasingly focused on the importance of the role of the nucleus instead of on the protoplasm in understanding life processes. This might also be a better description for the lasting contributions of Schleiden and Schwann, since, as has already been alluded to, the first tenets of cell theory may be in need of revision.

SCHLEIDEN AND SCHWANN

Matthias Schleiden was born in 1804 in Hamburg, Germany. He was always interested in botany but initially studied law, perhaps pressured by his wealthy family. He became a barrister in 1828, but soon became not only dissatisfied but deeply depressed, and attempted suicide. In 1831 he returned to college to pursue his true interests, botany and medicine, and eventually became a professor of botany at Jena University. What particularly interested Schleiden was the microscopic study of plant development. He helped train some of the leading botanists of the century, including Carl Nägeli (1817–91) and Wilhelm Hofmeister (1824–77). Schleiden recruited the zoologist Theodor Schwann for cytological research. He also encouraged the young optician Carl

Zeiss to start an optical firm and gave him enough orders to ensure the firm's success. Zeiss's contribution to building better microscopes was pivotal to the advances made in the life sciences in the nineteenth and twentieth centuries. To this day Zeiss's company remains a leader in high-quality optical instruments from microscopes to telescopes. Schleiden was impressed with the work of Robert Brown (1773–1838), who had identified the nucleus in many cells and clearly thought it had an important role. The nucleus had been observed in cells by a variety of people, but Schleiden emphasized its importance in the generation of the cell. Schleiden renamed it the "cytoblast," and doing so implied that this organelle was the structure from which the cell arises.

A key tenet of modern cell theory is the idea that cells only arise from preexisting cells. However, this aspect was not part of the initial formulation by Schleiden and Schwann. In fact, such an idea was repugnant to Schleiden, who wanted everything to be explained in strictly physiochemical terms. To claim that cells came from preexisting cells seemed to imply preformation, an idea that was decidedly out of favor by that time. Instead, Schleiden argued for an epigenetic approach and in 1838 proposed a theory of "cell-free formation." From the slime-like substance inside the cells, granules arose, accumulating material that eventually formed nucleoli, which grew and became more sharply defined. The cytoblast (nucleus) emerged surrounding the nucleoli and continued to grow. Once the nucleus had reached its full size, a delicate transparent vesicle, which he described as a young cell, appeared on its surface. In his detailed description, he claimed that the cytoblast became enclosed within the cell wall of plant cells. He thought that new nuclei might also form within existing cells, crystallizing from the formless fluid. For the next two decades the theory of cell-free formation was highly debated and eventually shown to be wrong. Nevertheless, this was a concrete, detailed theory that could be tested, thus advancing cytological studies to definitively answer the question of cell origination. Most important, Schleiden maintained that the basic unit of all plants was the cell. All the various shapes and sizes of the various structural elements of the plant were either cells or products of cells. What is particularly interesting about Schleiden's view is that, although he wanted cell origination to be explained in strictly physiochemical terms, at the same time he was actually a vitalist. Although not explicitly stated, he thought that some sort of vitalistic force animated the granules, causing them to aggregate and eventually give rise to a new cell. Thus, each generation a new cell was created *de novo*. One can see a parallel of Schleiden's view in the idea of spontaneous generation. However, eventually cell theory played an important role in ending the debates over spontaneous generation, but *not* in Schleiden's original formulation of it. Crucial to extending cell theory was establishing that cells were the basic structural units of animals as well as plants. Schwann was impressed with Schleiden's work and set out to prove just that.

Theodor Schwann was born in 1810 in Neuss, Germany. He studied medicine in several universities and after graduating in 1834 accepted a position in an anatomy museum in Berlin, assisting the physiologist Johannes Müller. While there he isolated the first digestive enzyme that had been prepared from animal tissue and named it "pepsin." Eventually he was appointed professor of anatomy at the University of Leuven, Belgium. There, he discovered that yeast were living organisms responsible for the fermentation of starch and sugars and coined the word "metabolism." He also made many important discoveries concerning muscles and nerves, including the myelin sheath that covered the peripheral axons known as Schwann cells. For our story, however, Schwann's most important research was in embryology. By following the embryonic development of a single animal cell into a complete organism, he showed that all tissues, even something like mature bone, which shows no evidence of being made up of cells, can trace their origin back to cells. However, he differed from Schleiden in that he thought new cells formed from the aggregation of materials that were *outside* of rather than *inside* the cell. Nevertheless, both men thought that the cells were truly new, not inherited or preexistent in any form. Instead, they thought cells were formed by a process of crystallization. Thus, the research of both men provided additional evidence for the fundamental unity of life. In 1839 Schwann summarized their findings in *Microscopical Researches into Accordance in the Structure and the Growth of Animals and Plants*. He maintained, first, that the cell is an organism and all living organisms are made of one or more cells. Second, the cell is the fundamental unit of both plants and animals. All of life's diversity has been generated by different arrangements and combinations of cells.

One can't overemphasize the importance of the discovery of cells in our quest to understand "what is life?" Yet at the same time, from a

purely technical point of view, the discovery was the inevitable cons-
quence of the rapid improvements in light microscopy. Nevertheless,
cell theory provided a unifying and ordered scheme for explaining
the organization of both animal and plant tissues. One of the marks
of a good theory is it provides a research program to further understanding and, more often than not, this is because various aspects of
the theory continue to be debated. Cell theory was no exception. The
basic two tenets of cell theory – that all organisms are made of one or
more cells and that the cell is the smallest unit of life – were generally
accepted by the mid-nineteenth century. However, that does not mean
that there wasn't controversy. Some people still challenged the exact
definition of the cell. Did it include a membrane and/or cell wall to
enclose it? In particular, a lot of confusion surrounded the origin of
the cell. Did it arise *de novo*, that is, from scratch, or did it come from
a preexisting cell? Answering this question gave rise to the third and
essential tenet of modern cell theory: cells only arise from preexisting
cells. However, it was not part of Schleiden's and Schwann's original
description. Furthermore, in part due to the prestige of these two men,
it was some time before their idea of cell-free formation (generating
cells from non-cellular material) would be discredited. While Louis
Pasteur is usually credited with doing the definitive experiment that
ended the debate over spontaneous generation, the importance of
observing microscopically that new cells were the product of cell division was absolutely vital (!) to the settling of this controversy.

REMAK AND VIRCHOW

Most histories generally credit Rudolf Virchow (1821–1902) with for-
mulating the last tenet of modern cell theory, that all cells come from
preexisting cells. However, he relied heavily on and confirmed the earlier research of Robert Remak. Several cytologists, including Heinrich
Rathke, Carl Bergmann, and Karl Reichart, were having trouble with the
idea of cell-free formation. However, the person most responsible for
discrediting the theory of cell-free formation, which ironically Schwann
considered the most important aspect of his contributions to cell theory, was Remak. Definitively disproving cell-free formation illustrates a
theme that repeats itself again and again in the history of science. The

ews of someone who is considered an authority are very hard to displace. Sociologist Robert Merton coined the phrase "Matthew Effect" in science based on a verse from the New Testament (Matthew 25:29): "For to every one that has shall be given, and he shall have abundance: but from him that has not shall be taken away even that which he has."[1] Schwann's histology was meticulous, and since he described his observations in such detail, others were reluctant to criticize them. When other researchers' findings contradicted his theory of cell-free formation, rather than disagree they tried to find a way to accommodate their findings. For example, Reichart observed the furrowing of cells in the early stages of development, but rather than asserting that this was actually the beginning stage of division that resulted in two new cells, he claimed that it was merely a mechanism for the mother cell to release a brood of new cells from the membrane that had already been formed inside the mother cell. Remak demonstrated that Reichart's interpretation was incorrect and, moreover, no process of crystallization of materials occurred as Schwann had claimed.

Remak is another scientist whose work was underappreciated in his own time, and remains so even today in regard to cell theory. Remak was born in Posen, Poland, which was under Prussian rule at the time. There he received his early education. He lived in Berlin for his professional life, but he remained loyal to his Polish identity. Being Jewish as well as being involved with several liberal organizations severely interfered with his academic career. Remak is best known for his many contributions to neurology. He was the first to describe unmyelinated nerve fibers (sympathetic nerves), now called Remak's fibers. He demonstrated that the gray color of these fibers resulted from a lack of a fatty white myelin sheath. He also discovered that the fibers from motor neurons in the spinal cord continued without interruption into the anterior roots and peripheral nerves. Later, Purkyně named these connections "axis cylinders."After receiving his doctorate, Remak continued neurological investigations in Müller's lab. In 1839 he described ganglion cells in the right atrium of the heart and related them to the sympathetic nervous system. They are still called "Remak's ganglion" today. In 1844, he was the first to demonstrate that the cerebral cortex consists of six layers. These many important discoveries should

---

1 Merton, "The Matthew Effect in Science," 56–63.

have ensured him a professorship. However, the antisemitism of the Prussians meant that Jews needed to convert to Christianity to advance. Remak refused to be baptized, and as a result he never obtained a professorship. With the necessity of earning a living, he became an associate in the clinic of Johann Schönlein at Charité Hospital and also had his own clinical practice. Finally, after years of effort and letters of support from Schönlein and the noted Prussian geographer and naturalist Alexander von Humboldt, he was offered an unpaid (!) extramural lectureship in 1847. In 1859 he was promoted to assistant professor, but this was a position far lower than he deserved considering his many outstanding achievements. His bitterness at how he had been treated was apparent when he dedicated his *Investigations on the Development of Vertebrates* (1855) to von Humboldt, writing with "eternal gratitude for his support in a life frustrated by religious and political prejudice."[2]

Just as with Purkyně, who is best known for his contributions to neurology, our interest in Remak concerns his research on the cell, which was an outgrowth of his interest in embryology. In 1817 Christian Pander had identified three germ layers in the developing chick embryo, and in 1828 von Baer applied the concept of germ layers to all developing vertebrates. In 1849 Huxley published a major article on the family of Medusae, or jellyfish. Huxley thought that the layers he identified in the adult jellyfish corresponded to Pander's germ layers in the chick embryo. Applying von Baer's techniques to studying development, Huxley realized that a correlation existed between the body architecture of the adult jellyfish and the vertebrate embryo. Based on these findings, Huxley was attempting to integrate development (ontogeny) with a natural system of classification (phylogeny) that would show how **invertebrates** were related to **vertebrates**. Six years after the publication of Huxley's work, Remak refined the germ layer theory in two important ways. First, as a result of his detailed microscopic studies of embryonic chick development, he observed that all the germ layers were derived from the original single cell of the fertilized egg. Thus, Remak concluded, all cells originate from the division of preexisting cells, a conclusion that became central to cell theory. This was contrary to Schwann's cell-free formation theory. Second, Remak provided histological evidence for the existence of three distinct germ layers

---

2 Quoted in Harris, *The Birth of the Cell*, 129.

and traced the derivatives of each throughout chick development. As previously mentioned, the idea of *de novo* cell origination had striking parallels to the idea of spontaneous generation, something that did not escape Remak. He claimed it was his doubts about Schwann's theory that caused him to investigate the multiplication of red blood cell precursors in the chick embryo. Various researchers had identified the blastoderm, the very first clump of cells that formed on the top of the yolk of the developing egg. By a systematic use of various hardening agents, Remak was able to definitively show that it was the membrane of the egg cell that was dividing to generate the blastoderm. There was no *de novo* formation of these blastoderm cells in the extracellular fluid, nor did he observe any naked nuclei. Remak continued his investigations. Whether it was the cells of primitive embryonic muscle bundles or the cells that were the precursors to the vertebral column, cell multiplication was occurring by binary fission. This was the rule, not the exception, at whatever stage he observed in embryonic development.

In an 1852 paper Remak extended the significance of this finding to pathology:

> It can hardly now be disputed that pathological tissue formations are simply variants of normal embryological patterns of differentiation, and it is not probable that it is their prerogative to generate cells in the extracellular fluid.... I make bold to assert that pathological tissues are not any more than normal tissues, formed in extracellular cytoblastem, but are the progeny or products of normal tissues in the organism.[3]

The assertion that malignant tissues arise from normal ones is particularly striking on several accounts. Not only was it in direct opposition to Müller's theory of tumor formation, but it was the central theme of Virchow's *Die Cellular-Pathologie*, which brought him considerable fame and prestige. Virchow has gotten the primary credit for not only the third tenet of cell theory but also the idea of tumor formation, in spite of the fact that Remak had explicitly made the same claims years earlier.

In his book on vertebrate embryology, Remak summarized his own research and expanded his critique of Schleiden and Schwann. He

---

3   Robert Remak, quoted in Harris, *The Birth of the Cell,* 131.

pointed out that Schleiden's and Schwann's theories of cell-free formation were not the same, and that they were both wrong. Schleiden,
whose observations were in plant tissues, claimed that new cells were
formed from amorphous material *inside* the cell. However, Schwann
could not confirm that in animal cells and instead argued that new cells
formed in the *inter*cellular spaces *between* cells. Yet as Remak pointed
out, this distinction implied a basic difference between plants and
animals rather than an underlying similarity. In addition, Remak had
no use for the process of crystallization. He argued that crystals and
cells were fundamentally different and, thus, how they formed bore
no relationship to each other. Remak had shown that early precursor
cells could differentiate into many different cell types. Detailing his
own research on how the embryonic layers differentiated, he argued
that the new cells were always being formed from material inside other
cells. All cellular development could be traced back to the fertilization
of the egg, which was a single cell. The appearance of multinucleate
cells might give the impression of cell-free formation, directed by the
enclosed nuclei, but he argued that appearances could be deceptive.
All cells had a nucleus and a membrane. Finally, he reiterated his position, advanced years earlier, that cell multiplication occurred by binary
fission only.

In light of this information, why has Virchow's name become syn-
onymous with the third component of cell theory? Both Virchow and
Remak were students and professional colleagues in Müller's lab. In
the early stages of their careers they discussed their experimental findings with each other. There is no doubt that Virchow was familiar with
Remak's work and that he accepted Remark's observations concerning
the formation of red blood cells. However, he was not initially sure that
all cells were formed by binary fission. In particular, he was not sure that
tumors formed that way, but rather that multinucleate cells might be
the centers of cell formation. But in 1855, in a lead article on cellular
pathology in the journal *Archiv* (which he founded), Virchow adopted
Remak's position in full. There were no free nuclei or extracellular
formation of cells. He saw that they were *not* forming by a crystallization process into the cytoblastema. Rather, life was continuous, one cell
giving rise to another: ***Omnis cellula a cellula*** (all cells from cells), ech-
oing William Harvey's "*ex ovo omnia*," all life from the egg. Yet Virchow
made no mention of Remak's work. Even in Virchow's time, this lack

of acknowledgment of Remak attracted criticism, and today it would be considered a significant breach of ethical standards. In the preface of Virchow's famous *Die Cellular-Pathologie*, he claimed that his article was an editorial and not a full scientific paper, and thus it was not necessary to acknowledge historical precedence, particularly of every small discovery. However, Remak's embryological findings, which Virchow made extensive use of in this book, hardly constituted a small discovery.

In examining the historical record, various authors have accused Virchow of everything from outright plagiarism to deliberately choosing to disseminate his views by way of the editorial to avoid mentioning his predecessors, particularly Remak. However, Henry Harris has argued that Virchow chose the editorial because it would have a much bigger impact, reaching a much larger audience, especially among doctors and pathologists, many of whom were not experimentalists.[4] Virchow was an outstanding lecturer as well as writer, and *Cellular-Pathologie* was an immediate success. The book was quickly translated into other European languages. There is no doubt that the rapid spread of Remak's ideas was due not to his own publications of his detailed observations, but rather to Virchow. In addition, Virchow's use of the phrase *Omnis cellula e cellula* was a short pithy denial of spontaneous generation, and remains a phrase that every biology student still learns today. By Virchow's time spontaneous generation had few followers, but this did not necessarily prove that cells all arose by binary fission. This was a critical addition to the cell theory, displacing Schwann's idea of cell-free formation. Remak certainly did not get the recognition he deserved, but *Cellular-Pathologie* was more than just a presentation of his work. Virchow included a great number of observations, some his own but also those from many other researchers. He brought them together to emphasize that all tissues, plant and animal alike, came from cells, and that cell multiplication occurred by binary fission only. Just as Schultze's ideas about the importance of the protoplasm received a much broader hearing due to Huxley's lecture and essay "The Physical Basis of Life," so did Virchow help spread the ideas of Remak. Nevertheless, there was an important difference. Huxley always acknowledged the work of other researchers. Indeed, the fact that he wrote so many popular essays bringing attention to many other

---

4 Harris, *The Birth of the Cell*, 135.

people's work, most notably Darwin's, has meant that the quality of his own significant body of original research has been underappreciated.

Just as Huxley was much more than a popularizer, Virchow was definitely more than a disseminator of other people's research, and his reputation is well deserved. Nevertheless, his relationship to Remak is problematic and raises questions for a historian of this period. Both Remak and Virchow had applied to the University of Berlin for the newly created Chair of Pathological Anatomy and General Pathology and Therapeutics. There was virtually no chance that Remak would be appointed because of his Jewish ancestry. The medical faculty recommended Virchow as their first choice, but Remak as their second. This is a testimony to the quality of Remak's research because, unlike Virchow, he actually had little experience in formal pathological anatomy. Virchow had little to worry about, but he wrote letters to several people to exert their influence to support his candidacy. Virchow also seemed to have believed in some sort of Jewish conspiracy theory and thought that von Humboldt's great support for Remak was due to his philosemitism. While Virchow's public views appeared to have been quite liberal, when it came to his own career he apparently was a victim to irrational bigotry. Even many years later, when Virchow's own status and academic position were totally secure, he continued to oppose any moves that would have advanced Remak's career.[5]

Remak indisputably deserves much more credit than he has received for the idea that all cells arise from other cells by means of binary fission. Nevertheless, there is also no doubt that Virchow played an essential role in the widespread acceptance of this idea. With this third tenet in place, cell theory was on its way to becoming a major unifying principle for biology. Although most researchers no longer believed in spontaneous generation, this third tenet helped counter the vitalistic ideas that remained prevalent in trying to understand this quality of "life." Instead, more and more experimentalists thought that the answer to the question "what is life?" would be found in understanding cell division. Remember, the mark of a good scientific theory is that it provides a research program for generating new knowledge. In this case, the recognition that the egg was a cell fully capable of developing into a

---

5 Harris, *The Birth of the Cell*, 132–37.

complete organism brought about a new question: was it the nucleus or the cytoplasm that was primarily directing development?

THE ROLE OF THE NUCLEUS

In the 1870s, again greatly aided by the continual improvement of microscopes, staining, and fixing techniques, cytologists were able to identify more and more structures in the cell and also to observe differences between the cells. The nucleus attracted the most attention, but other structures were also identified, and it is not surprising that researchers wanted to know what role all of these entities played in development. Biologists more in the tradition of Huxley, rather than focusing on smaller and smaller parts, instead investigated the role the cell played in the whole organism. Differences observed between cells might be the key to understanding what was known about heredity and evolution. Within this research program some investigators focused on the whole cell, while others focused on the nucleus, tracking the fate of a single fertilized cell as it divided and differentiated into many cell types.

We can see from his descriptions that Leeuwenhoek had undoubtedly observed the nucleus as had many others, but they had no idea what its function was. The first person to bring serious attention to it was the Scottish botanist Robert Brown (1773–1858). While investigating the fertilization mechanisms of plants in the *orchidaceae* (orchid) and *asclepiadaceae* (milkweed) families, he identified the structure and named it the cell nucleus in a paper that he read before the Linnaean Society in 1831. In this paper, he referenced the work of Franz Bauer, who had also observed the nucleus in other species of plants and had made beautiful detailed drawings of the cell.

Brown was one of the most important botanists of the period. Like many other distinguished naturalists of the time period, from Humboldt before him and Huxley and Darwin after him, he came to prominence as a result of his research from exploratory voyages. In 1801 he was the ship's naturalist on the *Investigator* for a surveying voyage along the northern and southern coasts of Australia. When he returned to England in 1805, he devoted himself to classifying the approximately 3,900 species he had collected, most of them new to science. He was an outstanding microscopist, and his observations played an important role in classification, as well as what would eventually become cell theory. He showed the fundamental differences between gymnosperms (plants, such as conifers, whose seeds are not enclosed in an ovary) and angiosperms (whose seeds are enclosed in an ovary and includes most trees, flowering plants, shrubs, and grasses). Brown may have been the first person to have observed cytoplasmic streaming within cells, when in 1827 he observed that the small particles ejected from pollen grains showed a continual jittery motion. He also observed the same phenomenon in inorganic matter. Although no one at the time had a theory to explain this motion, he recognized that it was an important property of all small particles suspended in a fluid. It was called **Brownian motion** and is still known by that term today.

Brown maintained that the nucleus was an important part of living cells and not just an artifact of the microscope. The argument concerning whether a given structure is "real" or is an artifact due to the processes of fixing and staining the cells has always been an important consideration in the development of microscopy, from the very first microscopes of Leeuwenhoek and Hooke all the way to the most powerful scanning electron microscopes of today. Is what being observed merely an artifact of the preservation process, and even if not, how does one interpret what one is seeing? Brown documented that the nucleus was found in the cells of many different types of organisms, both plants and animals. However, he did not extend his research into finding out what its function was or how it might fit into a more general theory of the role of the cell in development. This would be left to others. Specifically, studies on fertilization and the role the nucleus played in development began different lines of research that were spectacularly successful. Yet, somewhat paradoxically, these studies also eventually resulted in investigations into heredity and development becoming quite separate disciplines.

Observations of an embryo inside a fertilized hen egg predate Aristotle, but the idea that the embryo was the product of cell development was essentially unimaginable until the acceptance of cell theory. Furthermore, one had to recognize that both the egg and the sperm were indeed cells, each containing a nucleus. As early as 1841, von Kölliker's microscopic investigation of the spermatozoa of invertebrates observed that they originated in sperm-producing cells, and he thus concluded that they were cellular in nature. In 1871 Oscar Hertwig (1849–1922) observed that spermatozoa of the sea urchin fused with the egg nucleus five to ten minutes after they entered the egg. The sea urchin was widely used for the study of development because its eggs are transparent. Hertwig saw that only one spermatozoan fertilized the egg, and that once it had penetrated the egg, the egg secreted a transparent membrane (vitelline membrane) that blocked any other spermatozoa from entering. In 1876 the Swiss biologist Hermann Fol (1845–92) confirmed Hertwig's observations with further studies and definitively proved that fertilization was the fusion of an egg and a sperm cell. In his case he observed the penetration of a single sperm into the ovum of a starfish. Yet Fol initially denied that the nucleus had any structural continuity between generations or that it played any important role in heredity. In 1888 Theodor Boveri (1862–1915) discovered an organelle near the nucleus and named it the "**centrosome.**" He realized its importance to cell division, referring to it as the "true division-organ of the cell." However, some cells didn't seem to have them, and in others it seemed to disappear. As Fol continued his investigations in the late 1880s, he claimed that centrosomes were permanent structures, each parent contributing either two or one that divided right after fertilization. For Fol it was the centrosome that exhibited continuity, and he began to think that it was the key to understanding heredity, not the nucleus itself. As our story unfolds, interest in the centrosome will wax and wane. Daniel Mazia will revisit it in the 1980s and make a bold suggestion of its importance. However, this was not the case in the 1870s and 1880s.

Many other researchers disagreed with Fol, including Walther Flemming (1843–1905), Edouard van Beneden (1846–1910), and Eduard Strasburger (1844–1912). If fertilization was the result of the union of the egg and sperm, this meant that the nucleus did not just come into existence to facilitate cell division and then disappear. Rather, it maintained continuity through generations, and it must play an important role in heredity. When these men, along with others, actually observed mitosis and saw the nucleus divide during cell division, this bolstered the idea that it was the nucleus that contained the hereditary material and was responsible for directing development. There was considerable controversy around this idea.

DIVIDING THE NUCLEUS

Reaching consensus about what was actually happening when fertilization occurred did not happen overnight. In spite of improved microscopic techniques, including the development of the oil immersion lens, which allowed viewing the inside of the cell in much greater detail, researchers disagreed over many technical aspects of what was being observed. Various investigators, all highly skilled microscopists, often had patently different descriptions of what was happening in the early stages of embryonic development. Just as it took many years to put the theory of cell-free formation to rest, controversy also surrounded the details of how cell division actually occurred. With more and more researchers thinking that the nucleus contained the hereditary material, the question remained: how was it transmitted from generation to generation and how did cell differentiation occur? Since fertilization was the result of the fusion of the nuclei of the egg and sperm, these researchers thought the answer would lie in understanding the details of nuclear rather than cell division.

Remak probably had observed nuclear division as early as 1841, because he had clearly described the partitioning of a nucleated embryonic chick blood cell into two nucleated daughter cells. It would have been logical to conclude that the nucleus must have also divided. But remember that Schleiden and Schwann's idea of cell-free formation was still prevalent at this time. If a new cell forms *de novo*, it follows that the nucleus also forms *de novo*. This is exactly what Reichart had claimed. He observed that as the cell prepared to divide, the nuclear membrane disappeared and the nuclear contents dissolved. He claimed that new nuclei were synthesized *de novo* in the new daughter cells. He was correct about the dissolution of the membrane. His claim that the contents also disappeared was in part due to inadequate microscopic resolution, but undoubtedly was due more to trying to be compatible with the model of cell-free formation. By 1855 Remak's observations of nuclear division in the frog embryo, coupled with his inability to confirm Reichert's observations, convinced him that the nuclear contents did *not* dissolve. Rather there was continuity between mother and daughter cells. Nevertheless, it was hard to envision a model other than the actual direct partitioning of the nucleus, and thus he remained cautious in his conclusions. Did the nuclear membrane actually dissolve?

Remak suggested that first the nucleolus (a small, dense spherical structure in the nucleus) divided, then the nucleus, and finally the cell divided. But did he actually observe this? He interpreted his observations in a way that was consistent with his preconceived notions that binary fission was the mechanism by which new cells arose. Two other examples are worth mentioning and illustrate "seeing is believing," but at the same time also that one must seriously question what one is seeing. Carl Nägeli (1817–91) was another distinguished botanist who described the formation of the nucleus in a variety of ways. In some cases he claimed that it formed *de novo* in a cell, while the old nucleus was still intact and attached to the cell wall. In other cases he described the nucleus being partitioned essentially by a cell wall. In the spore mother cell of *Navicula striatula* (a species of diatom), he said the nucleus gave rise to two nuclei in the two daughter cells, while in *Anthoceros* (hornwort) he claimed that four nuclei formed from two nuclei. In both cases, he claimed to have observed the partition of one nucleus into two: "Now the first thing that happens is that this nucleus divides into two rounded nuclei, and then the formation of the two secondary mother cells takes place."[6] Although he made these detailed descriptions, he did not think the nucleus was essential to the cell as he described anucleated cells, and claimed that some species of plants had cells with no nuclei at all. Perhaps the most outstanding case of a description that was practically pure fiction was one made by Virchow, and is worth quoting at some length.

> The most common form of nuclear division takes place in the following manner. First a small constriction or groove is formed on one side of the usually rather oval nucleus. This groove gradually extends over the surface of the nucleus, and from it the partition wall then penetrates right through the interior of the nucleus. Sometimes one sees such grooves form simultaneously at two or more sites on the periphery of the nucleus which then divides into two or more subunits. To begin with the partition wall is always completely straight, but as the nuclear subunits grow and differentiate into discrete

---

6 Carl Nägeli, quoted in Harris, *The Birth of the Cell*, 139.

nuclei, their boundaries also round up: and eventually the new nuclei move apart.[7]

How could the esteemed Virchow with his considerable microscopic skills have written such a detailed description that we now know was blatantly incorrect and even had much contradictory evidence at the time? Perhaps because such a description provides a mechanism to support his view that cells arise from preexisting cells and by binary fission only. This is only one example of many that demonstrates how theory often directly influences how one interprets data – to the point that it actually shapes what one claims to have observed!

CHROMOSOMES

It is difficult to know who was the first person to actually see **chromosomes** as distinct entities during mitosis. It may have been Jakob Henle (1809–85), who in his 1841 textbook described the elongation of nuclei that then become long slender strands. But did he actually see chromosomes? In 1842 Nägeli also described structures that may have been chromosomes. However, in other writings, while he emphasized that cells of higher plants divided by binary fission, he did not totally reject the idea of *de novo* synthesis of the nucleus. Remak's 1858 paper contained plates that were clearly recognizable as anaphase, the stage in mitosis when the chromosomes are seen to be separating into two groups, and telophase, when they are totally separate. Yet he described them as shriveled nuclei and probably thought they were artifacts.

Another person who deserves serious consideration is the German botanist Wilhelm Hofmeister (1824–77). Working with various plant species in 1848, he described and illustrated the stages of mitosis that we recognize today (see figure 3.1). By staining the cells with iodine, he demonstrated that after the nuclear membrane dissolved, but before cell division occurred, material from the nuclei was still present; he referred to this as "**klumpen.**" He observed the klumpen attaching to an equatorial plate formed in the dividing cell (metaphase), the

---

7 Rudolf Virchow, quoted in Harris, *The Birth of the Cell,* 140. For original, see Virchow, *Archiv für Anatomie, Physiologie und Wissenschaftliche Medicin,* 89.

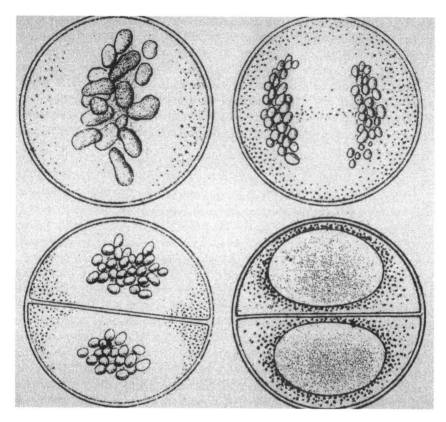

3.1 Hofmeister's illustration of the stages of mitosis: metaphase (upper left); anaphase (upper right); telophase (lower left); reconstitution of the daughter nuclei (lower right). Reproduced from K. Von Goebel, *Wilhelm Hofmeister: The Work and Life of a Nineteenth Century Botanist*, trans. H.M. Bower (London: Ray Society, 1926).

movement of the klumpen to opposite sides of the cell (anaphase), the formation of a nuclear membrane around two newly reconstituted nuclei (telophase), and the formation of a cell wall. In spite of this wonderful description, he had no idea of its function. However, he was convinced that it was not a random process, and he had clearly demonstrated that nuclear division was much more complex than just the simple partitioning that Remak and Nägeli envisioned. Improved staining methods had made it possible to see distinct chromosomes, but different researchers drew different conclusions. Strasburger maintained that nuclear division occurred transversely, that is, across the chromosomes. This resulted in different pieces and different materials from the same chromosome being partitioned into the two new nuclei. Van Beneden disagreed, claiming that the division was longitudinal, based on his observations of the movement of chromosomes during cell division. Flemming agreed with van Beneden. In a beautiful set of studies that he published from 1879 to 1881, Flemming described the different stages of mitosis. He argued that the division of the **chromatin** (the klumpen of Hofmeister) was responsible for cell division in all cells. By the end of the 1880s, consensus had emerged that the nucleus played an important and possibly *the* critical role in cell division. In this period, we also see new terminology that became standardized to describe the various parts of the cell. Strasburger originated the term "cytoplasm" to describe the contents inside the cell except for the nucleus and the term "nucleoplasm" to describe the contents of the nucleus. In 1888 Henrick Wilhelm Waldeyer gave the name "chromosome" to the chromatin.

In this same time period, several researchers, including Boveri, noticed that something special was happening in the production of germ cells (egg and sperm). The cell divided, giving rise to two cells, each with a complete set of chromosomes, but when those two cells divided again, the chromosomes did not duplicate themselves. Instead, they just separated. The end result was four cells, but each had only half the number of chromosomes found in the somatic cells. This process was initially referred to as reduction division and later called **meiosis**. This suggested that the chromosomes were important to heredity. Why? Because it would explain how the hereditary material could be transmitted from generation to generation and not result in a doubling of it every time an egg and sperm fused in a new generation. As we will see later, working out the details of meiosis was essential to identifying the chromosomes as carriers of hereditary information (see figure 3.2).

One of the problems with locating the hereditary factors on the chromosomes was that in the period before cell division, what is now known as interphase, the chromosomes seemed to disappear. How then could they contain the hereditary information? This is perhaps one of the reasons that Fol claimed that the centrosome was responsible for the hereditary information. However, using a light microscope to examine the eggs of the parasitic worm *Ascaris megalocephala,* which had only

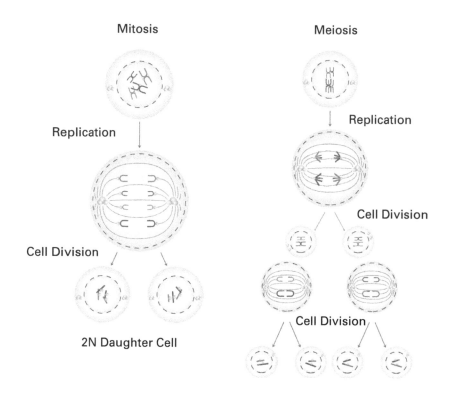

3.2 Schematic illustrating the differences between mitosis and meiosis.
iStock.com/snapgalleria

four chromosomes, Boveri showed that the chromosomes had the same shape and arrangement in the daughter cells before and after mitosis. This suggested that the chromosomes were highly organized structures. Since inheritance necessarily involves continuity from generation to generation, Boveri thought this made chromosomes good candidates for that function. He then switched his attention to sea urchin eggs. Upon careful examination he saw that the nuclei of the egg and sperm did not fuse as previously thought. Nevertheless, they each contributed equal numbers of chromosomes. Still, many people remained skeptical about chromosomes' role in both heredity and development. Are they all the same? Is a complete set needed for normal development and reproduction? In a series of experiments, Boveri produced eggs that had been fertilized by two sperm resulting in a **zygote** (fertilized egg). In the first cleavage, the zygote divided into four cells (referred to as **blastomeres**), not two, and each new cell had variable numbers of chromosomes. Only about 11 percent had the normal number of 36, and only those developed normally. Depending on which assortment of chromosomes a particular blastomere had, different kinds of abnormal embryos formed. In 1902 he published his results, writing that a "specific assortment of chromosomes is responsible for normal development, and this can mean only that the individual chromosomes possess different qualities."

Independently, Walter Sutton (1877–1916) came to the same conclusion working with grasshoppers. Sutton, a Kansas farm boy, became a student of C.E. McClung at the University of Kansas. The sea urchin was an ideal organism in which to study early development because of its transparent eggs. The initial reason for working with grasshoppers was more mundane for McClung. He took advantage of the plentiful supply of grasshoppers in Kansas for cytological study and founded a school of grasshopper cytologists. Sutton found a species, *Brachystola magna*, in which the male meiotic chromosomes were quite large and clear. He showed that the chromosomes occurred in distinct pairs that segregated at meiosis.

There was one other crucial finding that led to the location of the hereditary material on the chromosome as inescapable: the rediscovery and independent verification of Gregor Mendel's work. In the mid-nineteenth century, high in the Austrian alps, a Moravian friar by the name of Gregor Mendel (1822–84) was working out the laws of inheritance. Many historians have argued that was his primary concern, but I agree with historian Robert Olby that what primarily interested him was whether hybridization, as Carl Linnaeus (1707–78) had claimed, could lead to the production of new species.[8] Yet in investigating this question, Mendel laid the groundwork for the founding of a new discipline that would achieve spectacular successes in the twentieth century: genetics. Using the common garden pea plant, Mendel conducted a large number of experiments crossing and self-fertilizing thousands of plants. He discovered that when he crossed two true-breeding plants with different traits, for instance ones with yellow or green seeds, the first generation (F1) all resembled one parental type (for example, an F1 with all green seeds). He called that type **dominant**. However, when he crossed the F1 with

---

8 Olby, "Mendel, No Mendelian?", 53–72.

itself, the F2 ratio was 3:1, green to yellow. The yellow was not lost but only masked, and he called that trait **recessive**. By self-fertilizing the plants in the F3 generation, he found that the recessive plants bred true, one-third of the dominant ones bred true, and the other two-thirds gave a ratio of 3 green to 1 yellow. By doing several different crosses with several different kinds of traits, he found that the traits assorted independently, always giving approximately a 3:1 ratio of dominant to recessive. However, because he worked in relative obscurity, his research had virtually no impact until his 1865 paper was rediscovered and independently verified by three other researchers, Hugo de Vries (1848–1935), Erich von Tschermak (1871–1962), and Carl Correns (1864–1933).

The above account represents a rather simplified narrative of the significance of Mendel's work. Most genetic traits do not follow the simple dominant/recessive model. Mendel did several crosses that produced traits that seemed to exhibit blended inheritance, but by following the crosses through several generations, he observed that the traits were not lost. Most traits are also controlled by many genes, rather than single pairs. The seven traits he described that exhibited the clear dominant/ recessive dichotomy also segregated independently of each other. Was it just a lucky coincidence that they turned out to be located on seven different chromosomes? Building on Olby's idea that what primarily motivated Mendel was his interest in hybrids, the historian and philosopher Federico Di Trocchio suggests that Mendel probably started quite broadly, crossing all kinds of varieties differing in many different traits, and hoping that a pattern would eventually emerge that could be expressed in a mathematical formula. The pattern of independent assortment occurred precisely because Mendel probably disregarded traits that were linked on the same chromosome. In addition, the 3:1 ratio he observed did not depend on the dichotomy of dominant/recessive. There were many other examples of crosses that looked like examples of blended inheritance. For example, when he crossed a white-flowered plant with a red-flowered plant, the F1 generation all had pink flowers. Yet the F2 would yield the same 3:1 ratio, showing that Mendel's "factors" were not lost, but they also didn't depend on the concept of dominance to recessive.[9]

Sutton realized that the behavior of the chromosomes corresponded to Mendel's factors. He concluded his 1902 paper, "I may finally call

---

9 For further discussion of these points, see Allchin, *Sacred Bovines*, 163–68. See also Di Trocchio, "Mendel's Experiments," 485–519.

attention to the probability that the association of paternal and maternal chromosomes in pairs and their subsequent separation during the reducing division ... may constitute the physical basis of the Mendelian law of heredity."[10] The work of Boveri and Sutton seemed undeniable: the chromosomes were the carriers of the hereditary information. This became known as the Boveri–Sutton chromosome theory, or the chromosomal theory of inheritance. Yet there were still some skeptics, and paradoxically the most prominent one was the person, with the exception of Mendel, whose name we associate more than anyone else with what will become the newly created discipline of genetics: Thomas Hunt Morgan (1866–1945).

Morgan eventually won the Nobel prize for his work on *Drosophila* genetics and is considered one of the founding fathers of the discipline. However, Morgan did not begin his career as a geneticist, but rather as an embryologist. Following his career choices helps unravel the many different threads that characterize the complex relationship of heredity to development. Morgan in many ways can be considered a direct intellectual descendent of Huxley. Like Huxley, much of his early work was in morphology. Using comparative embryology and anatomy, cytology, and to some extent physiology, morphologists at this time were primarily interested in discovering the evolutionary relationships between organisms. However, their work was mainly descriptive. Thus their conclusions were highly speculative, as there was no direct way to test them. This was one of the reasons that Morgan became increasingly dissatisfied with morphology.

Several additional factors contributed to Morgan's disenchantment with morphology as a discipline. First, he thought that embryology had its own distinct set of problems and should not just be used as a tool to study evolutionary relationships. While much of Huxley's research also related to determining phylogenies, this was not Huxley's primary interest. Rather it was to understand development, and Morgan shared that interest. While Morgan is best remembered for his work in genetics, he never gave up his interest in development throughout his long and illustrious career. Morgan recognized that to investigate how development occurred required an experimental rather than a purely descriptive approach. Although he did his doctorate in morphology at

---

10 Sutton, "On the Morphology of the Chromosome Group," 39.

Johns Hopkins under the direction of W.K. Brooks, he also had a close association with H. Newell Martin, the head of the biology department. It is in his relationship with Martin that one can also see Huxley's influence on Morgan, albeit indirectly.

Huxley was the first person to advocate that physiology be taught as a distinct and *experimental* discipline, and also outside of medical schools. He complained to the Royal Commission about the lack of facilities to teach courses that included laboratories to train future scientists. With the passage of the Elementary School Act in 1870, he was given the opportunity to establish such a course at the School of Mines. It soon became quite famous. The first courses in elementary biology at Oxford, Cambridge, and Johns Hopkins universities were modeled on it. The now standard inclusion of laboratory practice in schools, even in primary schools, can be traced directly back to Huxley. Perhaps Huxley's most famous protégé was Michael Foster (1836–1907), who became the first professor of physiology at the University of Cambridge, a position that was created in part due to Huxley's lobbying efforts. Foster established Cambridge as a world-class center for experimental research in physiology, and it became a model for teaching biology throughout Britain and then the United States. (It was at Cambridge that Watson and Crick discovered the structure of DNA.) Martin introduced experimental teaching laboratories at Hopkins, regarding physiology as the queen of the sciences with embryology its servant. Huxley and Morgan may not have thought that morphology was the servant of physiology; nevertheless, they both thought that the interaction of molecules according to the laws of physics and chemistry was what governed development and, therefore, was responsible for the generation of form. This reflected a physiological rather than a morphological view of life's ultimate organization.

Morgan's eventual crossover to genetics was still strongly informed by embryology, but in particular by experimental embryology. After obtaining his PhD in zoology, he met Hans Driesch (1867–1941) at the Marine Biology Station in Naples, which further convinced him of the importance of experiment. While most experimental embryologists followed Boveri, who maintained that the nucleus was the site that contained the hereditary information, Driesch followed the German biologists Wilhelm Roux (1850–1924) and August Weismann (1834–1914). They had argued that development was mosaic, meaning that cell differentiation resulted from the qualitative parceling out of some kind of physical material during cell division. Thus, tissues and organs became different from one another because the cells they were derived from were qualitatively different from one another. Driesch's research, however, suggested that this difference came about by the relationship of the parts to one another and to the whole organism. The environment played an essential role in how development proceeded. Morgan also agreed, at least initially, with Driesch that the hereditary factors were located in the cytoplasm. This was based on research he had done with Driesch in 1894 in which they removed the cytoplasm from uncleaved centophore (comb jelly) eggs. The result were defective embryos. Morgan asserted there is "no escape from the conclusion that in the protoplasm and not in the nucleus lies the differentiating power of the early stages of development."[11] We will examine this research along with that of Roux and Weismann in more detail in the next chapter.

While these debates concerned development from the egg, Morgan recognized that the same issues also applied to regeneration. Organisms such as planaria or the hydra could regenerate the entire organism from just a part. The cells changed both their shape and function to regenerate the missing parts. This was a serious problem for the mosaic theory, since Roux and Weismann were claiming that some physical quantity was being divided up qualitatively as development proceeded. How could a part de-differentiate and then reform if it was missing material that was initially present in the egg or the blastomere? Morgan realized that some new terminology was needed to clarify different regenerative processes. For a salamander to regenerate a limb, both cellular proliferation and a rearrangement of cells occurred. These processes often occurred in conjunction but were, nevertheless, quantitatively distinct. He labeled the first process "**epimorphosis**" and the second "**morphallaxis**." Morphallaxis referred to those cases in which regeneration resulted from the remodeling of existing material without cellular proliferation, such as regeneration in hydra. Morgan had no problem with morphallaxis because, like Huxley, he had a physiological view of development that was independent of cell boundaries. The organism as a whole would sense a change in the shape and compensate. For Morgan, the overall organization of the organism was absolutely

---

11   Morgan, *The Frog's Egg*, 121. In Gilbert, "Embryological Origins," 316.

essential to understanding development. At the same time, more and more evidence was mounting in favor of Boveri's view that the chromosomes located in the nucleus were strongly implicated in development.

Edmund Beecher Wilson (1856–1939) and Nettie Stevens (1861–1912) demonstrated a critical correlation between nuclear chromosomes and organismal development in 1905. They showed that XO (containing only 1 rather than 2 sex chromosomes) or XY embryos became male and XX embryos became female. The sex of the organism appeared to be determined by particular chromosomes. Yet Morgan was skeptical. Again, like Huxley, he was a thoroughgoing epigeneticist, and the Mendelian and chromosomal theories of inheritance smacked of preformation. In common with earlier preformation theories, these factors referred to preexisting particles. In spite of his doubts, Morgan started massive breeding experiments with the fruit fly *Drosophila melanogaster* (his lab became affectionately known as the fly lab), eventually creating literally hundreds of different mutants in eye color and various other traits. In 1910 he observed that when a white-eyed male was crossed with the normal red-eyed female, the F1 generation was all red-eyed. However, as he continued his experiments, he noticed that the white-eyed flies were always male. He interpreted this result to mean that a nonsexual trait, white eye, appeared to be physically linked to the X chromosome, although he did not explicitly make that claim. In fact, a year earlier he had published a paper highly critical of the gene concept of inheritance, claiming the Mendelian "factors" were hypothetical. Yet here was evidence that suggested the factors were real, located on the chromosome, and could be experimentally manipulated. Morgan finally put to rest the controversy surrounding the cytoplasmic theory of inheritance in 1915 (actually not really, as we will see in later chapters) when he published his monograph *The Mendelian Basis of Heredity*, which summarized the results of his and his students' work. In it he lays out a detailed and comprehensive argument that chromosomes were the material basis for inheritance. Just as the sea urchin became the primary organism of choice to study development, the fruit fly became the main organism for genetic experiments. Morgan and his students went on to identify enormous numbers of mutants, and the field of genetics rapidly progressed.[12]

---

12 See Gilbert, "Embryological Origins," 307–51.

It is ironic that Morgan, an embryologist who initially disputed the gene theory of inheritance, became the person most responsible for the split between the study of development and heredity. The field of genetics became much more narrow in its focus, occupying itself primarily with the problem of transmission, that is, how the hereditary material was transmitted from one generation to another. In doing so, embryology was tasked with the much more difficult problem of understanding Huxley's question: how does form come into being? The assumptions that genes were the units of heredity, were located on the chromosomes, and were also involved with directing development did not negate the idea that something important was happening within the cytoplasm that was also critical to development. How did a cell differentiate and transform into different cell types, forming tissues and organs, and eventually give rise to a fern, a whale, or a human? These were the kinds of questions that remained of significant interest to Morgan. He still thought heredity and development were intimately interconnected. He recognized that regeneration was fundamentally a process that returned cells to their embryonic state. Thus, some component of the hereditary information or factors must be turned on or off at different periods of the organism's life, which allowed for differentiation and de-differentiation. However, when he wrote his book *Regeneration* in 1901, he had no idea what those factors were. Just describing what was happening in regeneration or development was not an explanation. Like Huxley, he thought development was a process that followed the laws of chemistry and physics. It was not just an unfolding of preexisting structures. Thus, even when the evidence that chromosomes were involved in sex determination seemed to be fairly compelling, for Morgan they were just indicators of underlying processes. They were not the process itself. He would eventually change his view about the role of chromosomes, but until 1910 he claimed that a fertilized egg might inherit a potential for maleness or femaleness. However, that potential was only realized through the process of cell differentiation and organogenesis. At the present state of knowledge, however, he felt that the problems of development were essentially unsolvable, and he decided to devote himself to further the understanding of the laws of heredity. This was something that could be tackled experimentally and quantitatively. The Nobel prize–winning immunologist Peter Medawar famously said, "Science is the art of the soluble." Morgan was as successful as he was in part because he devoted himself to problems that were solvable. Untangling the laws of inheritance was intrinsically a much simpler problem than understanding development.

With its spectacular success in its own rather narrow focus, genetics also eventually comes to dominate the thinking about evolution. The nucleus and the action of chromosomes moved to center stage. The protoplasmic theory of life continued to have its adherents, but it became an increasingly minority position. Nevertheless, embryologists – perhaps not explicitly, but certainly implicitly – acknowledged Huxley's caution about the limitations of cell theory. Embracing a more holistic perspective, they recognized that development involved cells interacting and communicating with one another, as well as the epigenetic perspective that had shaped Huxley's critique of cell theory. Many of the ideas that Huxley articulated continued to inform embryological experiments throughout the twentieth century. The next chapter explores how the advent of cell theory informed and in so doing also shaped the kinds of questions that experimental morphologists chose to investigate.

CHAPTER FOUR

Cell Theory in Development

Heredity is today the central problem of biology ... but the mechanism of heredity can be studied best by the investigation of the germ cells and their development.

Edwin Conklin, "The Mechanism of Heredity," 1908

Without a structure in the egg to begin with, no formation of a complicated organism is conceivable.

Jacques Loeb, The Organism as a Whole, *1916*

THE PROBLEM OF DEVELOPMENT

Our story so far shows that until the end of the nineteenth century the study of heredity and development were not regarded as separate disciplines; rather they were aspects of the more general question of generation. All living organisms reproduced, and the offspring resembled their parents. To understand why cats always gave rise to cats, and frogs likewise did not give rise to cats, one had to study both heredity and development. It was not possible to understand one without the other. Yet, as alluded to in the previous chapter and as will be explored in greater depth in the next chapter, in the twentieth century a split occurred and the two disciplines became quite distinct. In unraveling how the chromosomes carried out their function, most researchers were remarkably unconcerned with what was happening in the cytoplasm, or the role that cells played within the organism. Thus, our story now shifts to exploring the role cell theory played in understanding development.

Although most biologists accepted that the cell was the smallest independent unit of life, such a claim posed certain problems in elucidating how a single cell eventually gives rise to a multicelled organism. If a cell has an independent existence, then what is its relationship to the whole organism? In other words, how does one define an individual? This also highlights another controversy: what role does the whole organism play in directing development? Would a holistic/organicist approach or a reductive/mechanistic approach be the most effective in trying to understand the relationship between the individual cells and the whole organism? Advocates of organicism or (w)**holism** argued that the whole is greater than the sum of its parts. Mechanists, however, maintained that a complete explanation of the whole organism can be found by understanding its component parts, ultimately down to the laws of physics and chemistry. The old debate between preformation and epigenesis was worth revisiting and reframing as a result of improved microscopy and a greater understanding of the role the nucleus played in development. The most basic problem of development was to understand how cells became specialized and differentiated as they divided. Until the late 1800s embryological studies were primarily descriptive, and certainly this was critical to advancing theories about how development occurred, but a growing number of researchers began to think that the problem of cell differentiation would never be solved solely by careful observation. A more interventionist approach was needed. Various ingenious experimental approaches were taken, but often gave conflicting results. Cells sometimes seemed to be acting independently of their neighboring cells, their fate being determined right from the beginning in the egg. At other times, how development proceeded appeared to be regulated by the organism as a whole. Yet how these experiments were designed as well as how they were interpreted was profoundly influenced by differing conceptions of the cell.

The German philosopher and historian Hans Blumenberg has suggested that two background metaphors have shaped conceptions of the cell. The first one is as human artifact, where cells are regarded as spaces or rooms encased in a wall. This is in line with how Hooke first described the cell. Cells were considered to be building blocks or stones. As more was learned about the internal structure of cells, they were described as chemical laboratories that built the organism. Artifacts are something we create and as such we can understand them, just like the machines that we build. Cells in this conception are regarded as machines. The second background metaphor is that of organism. Beginning in the nineteenth century, cells were considered to be elementary organisms. Even cells within multicelled organisms were thought of as similar to single-celled organisms such as amoebae and other protists. However, in animals and plants the cells were interacting and influencing one another. In this conception cells are also *social* organisms and, as they interact, new properties emerge that would not be possible if acting alone or in isolation. Understanding the nature of these interactions would be the key to understanding development, to explain how a cell differentiates into a neuron or connective tissue.[1] Both of these background metaphors can be seen in the research discussed in this chapter; however, the chapter emphasizes the experimental approaches that view the cell as a social organism.

WHAT IS AN INDIVIDUAL?

Underlying all studies in development was a fundamental question: what was the relationship of the parts to the whole organism? As Georges Canguilhem wrote, "The history of the concept of the cell is inseparable from the history of the concept of the individual."[2] In the first part of the nineteenth century, cells were often portrayed as the building blocks to make a more complex organism. However, with the articulation of cell theory, which claimed that cells were the smallest independent unit of life, many biologists regarded them as autonomous individuals. Indeed, Schleiden and Schwann often referred to them as little organisms. Nevertheless, Schleiden also recognized that

---

1 See Reynolds, *The Third Lens*, 4–5, 25.
2 Georges Canguilhem, *La Connaissance de la Vie* [The Knowledge of Life], 1969, quoted in Reynolds, "The Theory of the Cell State," 74.

the cell was part of a larger whole, remarking that "each cell leads a double life; an independent one, pertaining to its own development alone; and another incidental, in so far as it has become an integral part of a plant."[3] Nevertheless, both Schleiden and Schwann clearly espoused what Whitman referred to as the cell-standpoint. Schleiden continued by saying that "the entire plant appears to live only for and through the elementary organ [the cell]." Schwann affirmed this idea by extending it to the entire organic world: "Each cell is within certain limits an individual, an independent whole."[4] Virchow, around the same time, began describing cells as citizens of a cell-state or society of cells, living rich, social lives as they interacted with one another. For many investigators, this provided a useful analogy for how to think about organisms. Just as human society was hierarchically organized with a division of labor, so too was the organism, with its system of cells, tissues, and organs. As Virchow pointed out, the organism was made up of individual cells, but he also referred to "cell territories." Thus, to understand the individual cell one had to situate it in the larger social context of a particular organ or tissue. This analogy provided support to Huxley's claim that cells were not independent. If cells were truly autonomous individuals, then why would they need to specialize? What tied them together as an organism? At least initially, many researchers were not particularly interested in answering this question, sympathetic to Herbert Spencer's (1820–1903) views, who had described the animal body as a "commonwealth of monads." However, if cells were part of a cell society, this meant that they were not just little automatons, their fates determined right from the beginning. Instead, development of the embryo was the product of the social interaction of similar cells with each other as well as interacting with dissimilar cells. Nevertheless, whether the cell was regarded as autonomous or as part of a society of cells raised another question: how does one define an individual?

Huxley recognized that the question of individuality was indeed profound and did not have an obvious answer. The standard way of thinking about individuals was as a "single thing of the same kind." However, Huxley claimed that such a definition quickly led to

---

3 Matthias Schleiden, 1838, quoted in Whitman, "The Inadequacy of the Cell-Theory," 639.
4 Theodor Schwann, 1839, quoted in Whitman, "The Inadequacy of the Cell-Theory," 639.

absurdities when applying it to the biological world. He pointed out the difficulties in his 1852 address to the Royal Society. "On Animal Individuality" drew on his earlier research as well as the work of various other scientists. His analysis provides insight to his objections to cell theory and reflected his interest as a morphologist in understanding the nature of form. There were many kinds of individuality, of which he mentioned three. The first he claimed was arbitrary. It depended on how we viewed something and was thus temporary – for example, referring to the individuality of a landscape. The second depended on some kind of law of coexistence, and if altered materially it would be destroyed. A crystal ground into powder would lose its individuality. It was, however, the third kind of individuality that was defined by a law of succession, or a definite cycle, that was relevant to life forms. An individual beat of a pendulum consisted of the successive places as it moved from a state of rest to a state of rest again. Likewise, for an organism, an individual was "one beat of the pendulum of life" with birth and death being the two forms at rest. The different stages of the life cycle represented the different places of the swing of the pendulum. For most higher organisms, these stages were not separated, but passed imperceptibly from one to another.[5] The newborn kitten gradually became a cat, and the individual cat then was the sum of all these changes. However, this was not the case for many other organisms, which revealed the total inadequacy of the common definition of an individual. Nature often made sharp demarcations between the different stages in the life cycle of invertebrates as well as in the plant world. They exhibited what could be called "compound individuality." Most people would claim that the butterfly that emerged from the cocoon was a distinct individual from the caterpillar that it came from, and the vast majority of biologists also agreed. But not Huxley.

Huxley adopted William Carpenter's definition of a biological individual, which was not based on its independence but rather consisted of the entire product of a single fertilized ovum. In this definition, the caterpillar, the chrysalis, and the butterfly constituted a single individual. The caterpillar shedding its skin was merely the separating of one part from another, and the skin could not live independently. The life

---

5 Huxley, "Upon Animal Individuality."

cycles of other organisms were even more problematic. Many organisms exhibited an alternation of generations, particularly plants, where two or more different forms existed. One was generated asexually by budding off, while the other was the product of an egg and a sperm. Based on his research from the HMS *Rattlesnake*, Huxley invented the term "zooids" to describe parts that budded off from a "higher" individual but still had an independent existence. One such example was the salp, which had two different forms: *Salpa democratica* and *Salpa mucronate*. *S. democratica* grew a tube that then produced a chain of tiny buds. These buds then turned into a long chain of "individual" *S. mucronate*. The chain eventually separated, each one growing an egg. The egg detached and developed into a solitary *S. democratica*. Both these forms were highly organized and distinct. They were "like individuals and yet not individuals, in the sense that one of the higher animals is an individual." In these cases, rather than alteration of generations, he preferred to claim that the individual equaled the sum of its "zooids."

For Huxley, the difference between the caterpillar skin and *S. democratica* was one of degree. Likewise was the distinction between molting, fission, and budding.[6] He maintained that the term "zooid," however, was one of mere convenience and had "nothing to do with the concept of individuality itself." For Huxley, individuals existed in many different modes and these "modes pass insensibly one into the other in nature." While it was useful to distinguish them for purposes of classification, he, nevertheless, concluded that "[t]he individual animal is the sum of the phenomena presented by a single life: it is all those animal forms which proceed from a single egg taken together." In Huxley's emphasis on embryology in defining an individual, the influence of not only Carpenter but particularly von Baer is apparent. Von Baer had characterized development as proceeding from an unspecified homogeneous state, becoming more specialized and heterogeneous. Scientists along with philosophers of science have continued to debate the meaning of biological individuality (stay tuned for the final chapter!), which in turn shaped their research interests. Huxley's definition profoundly influenced biologists throughout the twentieth century in several ways. In spite of some of his most important work being done on invertebrates, his

---

6 See Elwick, *Styles of Reasoning in the British Life Sciences*, 131–38.

definition privileged higher organisms. This resulted in an increasing emphasis on understanding sexual reproduction and the transmission of heredity information.

For our story, however, this is not what is most important. By essentially bypassing the common criteria of independence as the defining characteristic, and instead defining an individual by its entire developmental history, Huxley's emphasis was on the whole organism. It didn't matter that cells were capable of an independent existence, because that is not how they functioned in a multicelled organism. Huxley maintained that thinking of them as independent, as cell theory implied, was counterproductive to understanding development. Huxley's views once again provide a window into the controversies surrounding morphogenesis, as scientists pondered what would be the best strategy to unravel this difficult problem of development – how to get beyond not just *describing* but *explaining* how an egg developed into the larva that grew into a caterpillar that spun a chrysalis from which a butterfly emerged.

WHAT IS A CELL?

Are cells autonomous individuals and thus plants and animals consist of aggregates of cells, or are they merely parts of an integrated and interconnected whole? The tension between these differing views continues to be an undercurrent in the research on development throughout the twentieth century. The anatomist Martin Heidenhain (1864–1949) attempted to reconcile these two different definitions of the cell. Heidenhain was not ready to abandon the cell concept and settle for the idea that protoplasm was the primary constituent of an organism and cell boundaries were incidental. At the same time, he thought that cell theory implied that cells were entirely independent and autonomous individuals within higher organisms. This was simply not true. He offered his own definition of a cell as "a limitable clump of living matter, with the morphological and physiological character of an elementary individual."[7] In support of the cell concept he cited several kinds of evidence. He drew parallels

---

7 See Reynolds, *The Third Lens*, 42.

between cells in tissues and unicellular plants and animals as well
as noting that individual cells can survive in tissue culture outside
of the body. White blood cells freely move about like autonomous
beings in the vertebrate body. Cells only arise from preexisting cells
as a result of cell division. At the same time, he recognized that in
higher animals and plants the individual cells were so subordinated
to the whole organism that they really functioned more like a part
than as an independent organism.

To further complicate the story, the status of the infusoria (later
classified as protists by Ernst Haeckel) was also contested. With the
advent of cell theory they were described as single-celled organisms
that had either plant- or animal-like qualities and sometimes both.
In the early twentieth century Clifford Dobell (1886–1949), who had
devoted his career to studying protists, maintained that they were
non-cellular or acellular in their organization. Dobell claimed that
only someone blinded by cell theory could fail to see this, and he
thought the theory should be abolished on the grounds of being
oversimplistic. The problem was not that one couldn't give a definition of a cell, but that it was being asked to be too many things. It
was an independent organism such as an amoeba, and it also was a
part of an organism as making up tissues. The fertilized egg was also
a single cell, but in this case it could be considered to be a potential
whole organism. How could it be all of these things? Dobell's critique
influenced many other scientists. For example, in his 1920 textbook
on cytology, Leonard Doncaster (1877–1920) claimed that the idea
of cells as discrete and independent units had essentially been abandoned by most researchers. Furthermore, many distinguished biologists maintained that the whole group classified as protists should
be regarded as non-cellular, and that "the word 'cell' is beginning
to lose its definite and precise significance." Doncaster suggested
instead that the cell should be regarded as "a convenient descriptive
term [rather] than as a fundamental concept of biology."[8] As will be
discussed in chapter 7, these critiques of the cell will find resonance
with some present-day researchers who claim that cell theory should
be revised.

---

8 Quoted in Reynolds, *The Third Lens*, 42–43.

HOLISTIC VS. REDUCTIONISTIC/MECHANISTIC APPROACHES TO DEVELOPMENT

While genetics was able to use a strictly reductive methodology to great success, this has not been the case for development. Trying to answer whether development was by epigenesis or preformation, undoubtedly the longest-running debate in embryology, reveals another underlying controversy that has shaped the investigations of "life." As mentioned at the opening of the chapter, what approach will be the most fruitful in unraveling how an organism develops from a single cell? Organicism or holism maintains that the whole is greater than the sum of its parts. Understanding the parts is dependent on understanding their relationship to the other parts and to the whole organism. Properties emerge in the whole organism that cannot be predicted by reducing it to understanding its component parts as mechanists argue. Many (but by no means all!) developmental biologists as well as historians of the field argue that some form of organicism has always shaped embryological studies. As they have written, organicism has gotten a "bad rap" because some of its advocates believed in a kind of vitalism and in later years was "keeping bad company" when it became associated with Nazi organicism and Marxist dialectical materialism.[9] In contrast, mechanism became conflated with materialism and the reductive methodology that had been so successful in the physical sciences. Why shouldn't the life sciences be equally amenable to such an approach? Before proceeding, however, it is useful to clarify a bit further what these terms mean.

The idea of mechanism and its usage in the life sciences is closely linked to the mechanical philosophy of the seventeenth century. Its roots, however, are far older and can be traced back to the atomism advocated by early Greek philosophers, particularly Epicurus (341–271 BCE). However, it was Rene Descartes (1596–1650) and Pierre Gassendi (1592–1655) who provided the first systematic and also the most influential accounts of the mechanical philosophy. Likewise, organicism can trace its roots back to the ancients, particularly Plato, who maintained that the universe and its parts can be thought of as an organic whole, literally claiming it was a living intelligent being. Just like

---

9 Gilbert and Sarkar, "Embracing Complexity," 4–5.

mechanism, however, it was in the eighteenth century that Immanuel Kant (1724–1804) provided a systemic explanation of organicism that later researchers in the life sciences adopted: "The first principle required for the notion of an object conceived as a natural purpose is that the parts, with respect to both form and being, are only possible through their relationship to the whole.... Secondly, it is required that the parts bind themselves mutually into the unity of a whole in such a way that they are mutually cause and effect of one another."[10] Thus, as Huxley argued in his critique of cell theory, the organism as a whole was informing and shaping how development occurred. This has been characterized as a "top-down" approach. Huxley has usually been regarded as a mechanist because he also maintained that life could be understood completely in terms of atoms and molecules, strictly obeying the laws of physics and chemistry. No immaterial or mystical forces were needed. Many embryologists agreed with Huxley, but this did not mean that a complete understanding of development would be found by just *reducing* it to understanding its component parts. Rather, certain properties at higher levels of complexity emerged as a result of the interaction of component parts. Yes, atoms combine to form molecules, which make up cells, which eventually give rise to tissues and organs, but understanding how they do so would not be found by just analyzing the component parts. In contrast, others, most prominently the distinguished physiologist Jacques Loeb (1859–1924), did not agree. In his influential *The Mechanistic Conception of Life* (1912), he argued that a complete understanding of development would come about when it was explained in terms of physiochemical mechanisms. Loeb was trying to rid biological research of any traces of vitalism. For Loeb and his followers, this "bottom-up approach" (from atoms and molecules to cells to tissues to organs) was completely adequate to explain how life arose from nonliving molecules. Yet many embryologists would agree with Oscar Hertwig (1849–1922), who proposed that organicism was the true middle ground between vitalism and reductivism: "The parts of the organism develop in relation to each other, that is, the development of the part is dependent on the development of the whole."[11] They instead made use of both bottom-up and top-down approaches as

---

10 Immanuel Kant, quoted in Gilbert and Sarkar, "Embracing Complexity," 3.
11 Oscar Hertwig, quoted in Gilbert and Sarkar, "Embracing Complexity," 3.

they tried to resolve whether the embryo developed via preformation or epigenesis.

THE PREFORMATION/EPIGENESIS DILEMMA

As has been pointed out, preformation had many problems. It was hard to imagine how the entire future of a species could be incapsulated in the egg. In fact, to a modern reader the idea seems ludicrous. Nevertheless, this history also shows that the evidence was not as clear-cut as one might think. Our current understanding of development draws on both of these traditions. Certainly a little fully formed cat did not exist inside the egg of a pregnant cat. Yet epigenesis also had a fundamental problem that we are still unraveling today, and that is the question of differentiation. Even after the recognition that the egg was a single cell that was not just a structureless blob, how did the egg eventually form all the structures of the cat: its claws, its heart, its kidneys, its fur, its tail, its eyes? As the family cat sits purring contentedly by the fire, how did one cell give rise to a whole organism that was capable of expressing such an emotion? Observations, as primitive as they were, did indeed show that new structures appeared as development proceeded. The idea of a little homunculus was clearly untenable, but that did not mean that epigenesis did not have serious problems as well. Since offspring did resemble their parents, this suggested that some sort of information was being transferred from generation to generation. Research clearly demonstrated that the egg was highly organized. Not only did it contain a nucleus, but improvements in microscopy revealed that the cytoplasm also contained various other structures and was not homogeneous.

The discovery that chromosomes were the carriers of the hereditary information did not address the fundamental problem of how a single cell gave rise to many different types of cells. Questions still remained. Did the cytoplasm or the nucleus control development? Were the hereditary units divided up unequally in the daughter cells as development proceeded? Furthermore, chromosomes were in a sense preformed units that were inherited from the past. As we saw in the last chapter, this was one of the reasons Morgan was initially against the idea of the chromosomes and the nucleus directing development. It

brought to mind preformation. However, by 1934 he had swung completely to the other side, claiming that development was controlled by
genes on chromosomes, which were located in the nucleus. It would
seem that he no longer shared the views articulated by Huxley. Huxley,
however, had other intellectual heirs, some of whom will come into
direct conflict with Morgan as he confined his research to problems
of genetics rather than development. The kinds of issues that Huxley
articulated underlay the research done by a variety of researchers –
all who would make significant contributions to the emerging field of
experimental embryology.

Edmund B. Wilson quickly accepted the chromosomal theory of
inheritance and argued that development was directed by the nucleus;
nevertheless, he was deeply influenced by Huxley. Even if the nucleus
orchestrated development, Wilson also maintained, "the cell cannot
be regarded as an isolated and independent unit. The only unity is
that of the entire organism."[12] He claimed that as long as there was
continuity of the cells, they cannot be regarded as morphological individuals. Rather, they are how the organism as a whole divided up the
various physiological functions. In his classic book *The Cell in Heredity
and Development* (1900), Wilson wrote,

> in attempting to analyze the problems [of development] we
> must from the onset hold fast to the fact, on which Huxley
> insisted, that the wonderful formative energy of the germ is
> not impressed upon it from without, but is inherent in the
> egg, a heritage from the parental life of which it was origi-
> nally a part. The development of the embryo is nothing new.
> It involves no breach of continuity, and is but a continuation
> of the vital processes going on in the parental body.[13]

Yet how do the adult features lie latent in the germ cells and then
appear as development proceeds? "This is the final question that looms
in the background of every investigation of the cell." Although Wilson
wrote this in 1900, it had been the question on the minds of many

---

12 Wilson, "The Mosaic Theory of Development," vol. 2, 8.
13 Edmund Beecher Wilson, quoted in Moore, *Science as a Way of Knowing*, 444–45. For
    original see Wilson, *The Cell in Development and Inheritance*, 396–97.

biologists decades earlier, and would continue to confound researchers for decades to come.

DEVELOPMENT AND EVOLUTION

Huxley's own research in developmental morphology had been profoundly shaped by the embryologist Karl Ernst von Baer (1792–1876). Huxley had devoted much of his career to promoting, explaining, and defending Darwin's theory of evolution. However, as shown in chapter 2, he maintained that development was key to understanding evolution. At the end of his life, when asked to evaluate Darwin's position in the history of science, he wrote that von Baer "would run [Darwin] hard in both breadth of view and genius."[14] By examining the embryonic development of a variety of organisms, von Baer concluded that organisms could be grouped into four major types: the mollusca, articulate, radiate, and vertebrate. All animal forms had undergone some kind of differentiation and the further back one traced development, the more similar widely different animals appeared. He asked, "Are not all animals essentially similar at the commencement of their development – have they not all a common primary form?" He even suggested that at the earliest stages, the embryos of invertebrates and vertebrates would be indistinguishable. For von Baer the "history of development is the history of a gradually increasing differentiation of that which was at first homogeneous." His ideas were summed up in his four famous laws of development: (1) general characters of a large group of animals appear earlier in their embryos than the more special characters; (2) from the most general forms the less general are developed, until finally the most special arise; (3) every embryo of a given animal form, instead of passing through other forms, rather becomes separated from them; and (4) fundamentally therefore, the embryo of a higher animal form is never identical to any other animal form, but only to its embryo. While von Baer argued against transmutation because of this idea of distinct types, Darwin realized that embryology provided the strongest evidence for his theory of common descent. As he wrote in *The Origin*,

---

14 Thomas Huxley to George Romanes, 9 May 1892, in Huxley, *Life and Letters of Thomas Henry Huxley,* vol. 2, 242.

"All organic beings have been formed on two great laws – Unity of type and the Conditions of Existence."[15] Unity of type is often referred to as the archetype and refers to a basic plan or type that existed at different taxonomic levels, providing a model for the whole group. Furthermore, "Community of embryonic structure reveals community of descent." With the publication of *On the Origin of Species* in 1859, much of the discussion about development shifted to how it related to evolutionary theory.[16]

The word "evolution" has an interesting history that is also highly relevant to our story, coming from the Latin infinitive *evolvere*, meaning "to unfold or disclose." Evolution in the late 1600s was used to describe embryological development. Haller in 1744 applied the term to characterize the preformationist embryology of Jan Swammerdam. By the 1800s evolution became associated with this latter view, a kind of progressive embryological development. Naturalists also started to apply the ideas from embryology to the fossil record, and claimed that the development of species recapitulated the development of the individual, that is, the development of more complex organisms over time repeated the stages of embryonic development observed in an individual animal. Thus evolution came to mean *both* embryological and species progression. This idea was most fully developed in the research of the German biologist Ernst Haeckel (1834–1919).

Unlike von Baer, Haeckel was a strong proponent of Darwin's theory, and like Huxley in England, did a great deal to popularize evolutionary theory in Germany. Haeckel was also one of the first people to combine cell theory with the theory of evolution to try to find how multicellular organisms might have evolved from single-celled organisms that he called protists. The theory of the cell state was being taught throughout German universities, but Haeckel had learned it directly from Virchow. While Virchow had used the metaphor of the cell state to think about the relationship of cell parts to the whole organism, Haeckel used it in his thinking about how complex modern cell societies represented by animals evolved from more primitive kinds of cell societies.[17]

---

15 Darwin, *The Origin of Species*, 233.
16 For further discussion on these points, see Lyons, *Thomas Henry Huxley*, chapter 2.
17 Reynolds, *The Third Lens*, 28.

Haeckel was one of the foremost investigators of microscopic life in addition to being an advocate for evolution. He wrote and beautifully illustrated several volumes that were entirely about single-celled organisms as well as on marine invertebrates such as sponges, corals, and jellyfish. In these works we see how the idea of the cell state shaped his speculations. He claimed to have observed single-celled organisms that lacked a nucleus, which he named *Monera*, and claimed that they were the most primitive and ancient type of organism. He postulated that one such variety gradually gave up its solitary ways, forming simple colonies that were similar to present-day protist colonies. Comparing them to monastic communities, he wrote that they had a very simple social organization with a limited division of labor. But gradually the cells became more specialized and differentiated, each with its own specific task. Eventually organisms would evolve in which the cells were so interdependent that they could no longer survive on their own.[18]

By 1866 Haeckel had proposed a hierarchical theory of biological individuality that began with single cells, then organs made up of cells, and then single organisms that were multicelled. His classification also illustrated the difficulties that Huxley had pointed out in defining biological individuality. Haeckel recognized that colonial organisms such as the Portuguese man o' war contained a mixture of individual cells (Huxley's zooids), and a community of cells. He also made a distinction between cell states in which the individual cells had more autonomy and cell states in which the activity of the individual cells was more restricted and subject to a centralized system of control. Plants were an example of the first type, and he referred to them as cell republics. Animals illustrated the second type, and he claimed they were more like cell monarchies. In spite of Haeckel adopting the metaphor of the cell state, in which cells were interdependent, he nevertheless differed from Huxley in that he thought that cells were the primary physiological and anatomical unit. In addition, the cell was an important evolutionary unit in Haeckel's scheme.[19]

Haeckel and Huxley may have disagreed over the role of the cell in the organism, but they both thought that development and especially comparative embryology provided important evidence in favor

18 Reynolds, *The Third Lens*, 27–28.
19 Reynolds, *The Third Lens*, 26–29.

of evolution. However, they were the exceptions. Most biologists used findings from development, the concept of type, and the fossil record to argue *against* transmutation. For them, the fossil record was not evidence that one species actually evolved into another, but was just part of a wider developmental plan. It was Darwin's genius that he was able to link the findings from embryology to paleontology in support of his theory. As he wrote to Asa Gray, "embryology is to me by far the strongest single class of facts in favor of change of forms."[20] Darwin was thrilled when Fritz Müller showed that the *Nauplius* larvae had a structure common to all crustaceans, and Alexander Kowalevski observed that invertebrate tunicates shared the same larval structures as vertebrates. This did nothing less than unite the entire animal kingdom. It confirmed what Huxley was trying to demonstrate in his work on the Medusae: embryonic and larval forms showed common ancestries for diverse species. Darwin's ancient ancestors were the archetypes of various animal species. He believed that in these ancient animals the adult form and the embryo were similar. The archetype was in some degree embryonic and, therefore, capable of undergoing further development. This was the explanation for Darwin as to why embryos resembled ancient fossil forms. For von Baer, these resemblances were the necessary consequence from a common starting point by a single process of increasing specialization. Von Baer's third and fourth laws were specifically to distinguish his ideas from those of Haeckel's, who claimed that the ancestral stages of the adult were repeated in the embryonic stages of its descendants, forming a linear hierarchy. This is encapsulated in the phrase "ontogeny recapitulates phylogeny" and is referred to as Haeckel's biogenetic law, or **recapitulation**. Not only did von Baer's ideas predate Haeckel's, but for Darwin, the pattern in the fossil record was the result of descent from a common ancestor, with divergence and increasing specialization occurring over time. Although the variations from the general archetype were inherited, they usually did not make their appearance until late in development, while the embryonic stages remained unchanged.

Darwin's theory has gotten the lion's share of attention by historians of nineteenth-century biology, but this does not mean that important research was not going on in other fields. Darwin was able to draw on

---

20 Darwin, 10 September 1860, *Life and Letters of Charles Darwin*, vol. 2, 131.

a vast number of new findings from many different fields: exploratory voyages bringing back a wealth of organisms never seen before, fossil finds, and enhanced embryological observations made possible by improvements in microscopy. The basic tenets of cell theory also could not have come about without the tremendous improvement in microscopes and staining techniques.

To reiterate, cell theory, like evolution, is one of the few great unifying theories in biology. Von Baer's work was complementary to cell theory because it implied that to understand development, one had to trace the embryonic stages back to the very beginning, which was the single cell – the egg. All the information to make an organism was contained in the egg, and in that way von Baer could be considered a preformationist. Nevertheless, his work was the embodiment of an epigenetic perspective, since his research demonstrated that the parts of the organism gradually appeared as development proceeded. As is often the case in science, a deeper understanding eventually emerges that is a combination of previous ideas that were in opposition to each other. Catastrophism and uniformitarianism were opposing theories in geology about the history of the earth. Many histories claim that uniformitarianism turned out to be correct, but in reality our modern theory of earth history draws on elements from both theories. Likewise, our current understanding of development draws on elements from both epigenetic and preformationist traditions.

CELL DIFFERENTIATION AND ITS RELATIONSHIP TO DEVELOPMENT

A single fertilized egg develops into a complex organism containing literally billions of cells – cells that have become specialized. They then combine to form tissues, organs, and highly intricate structures. To what extent are these structures predetermined or preformed, and to what extent do they unfold in a stepwise manner that is influenced by the interaction between the parts as well as environmental factors both internal and external to the whole organism?

Wilhelm His (1831–1904) agreed with von Baer and recognized that if the germ cell did not contain preformed parts of the adult, whatever was responsible must nevertheless be present at the beginning of development. Once again, to a modern reader this hardly seems profound; indeed, it might appear quite obvious. Where else could the parts come from other than the zygote? However, His was suggesting more than that. He was claiming that the egg contained factors that were localized and were responsible for the development of the embryo. Development was not due to some immaterial vitalistic organizing force, but rather to some material substance. His had been trained in cytology by Remak, Virchow, and others. His improved microtome made possible the very thin serial sectioning of the embryo that would enable researchers to track minute changes in the nucleus and cytoplasm as development proceeded. He concurred with von Baer that one had to go back to the beginning, the fertilized egg, to understand development. Also, like von Baer, he disagreed with Haeckel's biogenetic law. For Haeckel, all organisms were essentially identical until the gastrula stage (the stage when the embryo is a hollow, cup-shaped structure having three layers of cells), and it was the historical evolutionary past of the organism that determined its development. His disagreed and was able to show that even at the earliest stages, embryos from different organisms were already different. Development occurred by essentially a mechanical process dependent on the physical conditions of the organism. From the very beginning the egg had prelocalized areas, which he referred to as organ-forming germ regions. These areas were in the cytoplasm, not the nucleus, and thus any study of embryonic development must recognize the importance of the cytoplasm. Although His did not discover any organ-forming germ regions or identify specific factors, this idea of prelocalization was an extremely useful hypothesis and suggested a variety of experimental approaches that were carried out by others. He referred to his approach as the physiology of development. Embryology up to now had been primarily descriptive. His's ideas provided the basis for a research program that would become more interventionist and known as analytical embryology or *Entwicklungsmechanik*.

His did not discount the importance of the nucleus or of heredity but, like Huxley, he argued that to understand development one had to look at the chemical and physical processes in the organism as a whole. Cells were in contact with with one another and influencing their behavior. Both men were morphologists; both were interested in discovering the laws that governed development or, as Huxley wrote, "how form came to be." Yet to understand the generation of form, one couldn't just study structure; one had to do physiological morphology. As mentioned, Huxley was one of the foremost advocates in the English-speaking world for the idea that the natural sciences be taught in schools. In addition, Huxley played a prominent role in making German research available to the British and Americans. Unlike most of his English colleagues, Huxley was fluent in German. The Germans were leaders in developmental research. In 1853 Huxley translated von Baer's fifth Scholium and part of the sixth Scholium of the *Entwicklelungsgeschichte der Thiere*. Huxley played a crucial role not only in raising various problems with cell theory, but in advocating a research program that pursued a physiological approach to understanding morphology.

Wilhelm Roux was one of the first embryologists to develop *Entwicklungsmechanik* into a full-fledged experimental program. Many of his ideas would need significant modification, and many of his experiments were defective. Nevertheless, he raised important questions and suggested an approach to answering them that was extremely useful. He founded and was the editor of the first important journal of analytical embryology, *Wilhelm Roux' Archiv für Entwicklungsmechanik*, which began in 1894 and still continues today. One of the first questions Roux asked was whether the egg needed a specific stimuli from the environment for development to proceed. Could the fertilized egg develop independently as a whole as well as its individual parts? Or was normal development dependent on environmental forces directing it, including cell-cell interactions? Botanists had long been describing the diverse effects of various environmental stimuli on growth and differentiation. Light, gravity, wind, temperature, and moisture affected virtually every aspect of plant growth, from the production of chlorophyll to the rate and pattern of growth to whether they kept their leaves or lost them. By constantly rotating frog embryos, Roux sought to eliminate any directional influence that gravity, heat, light, or magnetic force might be exerting on them. Since they all developed normally, he concluded that the embryo did not need a stimulus from an external agency and claimed that the development of the egg may be regarded as self-differentiation.

Roux then proceeded to investigate if self-differentiation was also true for the development of individual parts, and whether His's hypothesis that certain factors existed in particular parts of the egg were what determined how and where specific organs would develop. Although embryologists had been observing the early stages of development in a variety of organisms and had been describing them for many years, no one knew what the underlying mechanisms were. Was early development primarily mosaic or regulative, meaning were the parts of the egg already irrevocably determined on a specific developmental path, or were they subject to influences that allowed a certain amount of developmental plasticity? Roux discovered that the early development of frog embryos followed certain rules that might have broad applicability. Other researchers had also found that the embryos of bony fish and Ascidiacea or sea squirts followed the same rules. The first division resulted in the embryo being divided into two hemispheres, referred to as animal and **vegetable poles**. Generally the cells from the **animal pole** were smaller and divided much more rapidly. Early development usually resulted in the formation of a **blastula**, a hollow cavity formed by a single layer of cells. Roux observed that after the sperm entered the egg, a gray crescent formed 180° from where the sperm entered. The plane of the first cleavage was on the meridian of the entry point of the sperm, and the animal pole bisected the gray crescent. The gray crescent lasted at most a few cleavages, but from it a dorsal lip formed. The anterior-posterior axis formed in relation to the dorsal lip and determined what would become the **ventral** or front and the dorsal or back part of the embryo. When the neural fold formed the **blastopore**, the central cavity that formed in the first stages of development would be at the posterior end. What this means is that the plane of the very first cleavage divided the egg into right and left halves. Roux realized that this seeming invariant pattern suggested a way to investigate His's idea that certain factors existed in particular places in the egg and were responsible for the development of particular organs. If His was correct, after the first cleavage each cell would contain specific factors for only one half of the developing embryo. Alternatively, all the factors could be preserved in each cell division. By using a hot needle Roux succeeded in destroying one cell from the two-cell stage. In only about 20 percent of the operated-on eggs did the undamaged cell survive; the rest were completely destroyed. Roux collected more than 100 eggs that had one of the halves destroyed. In the case where the dead half remained attached, abnormal development proceeded in a particular

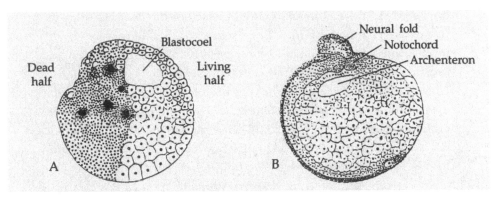

4.1 Roux's 1888 drawing of half embryos obtained after killing one cell of the two-cell stage of a frog embryo. In A the dead half remains. In B it has been sloughed off. Reproduced from W. Roux, "Beiträge zur Entwicklungsmechanik des Embryo. Ueber die künstliche Hervorbringung halber Embryonen durch Zerstörung einer der beiden ersten Furchungskugelin, sowie über die Nachentwicklung (Postgeneration) der fehlenden Köperhälfte" *Virchows Arch. Pathol. Anat. Physiol.* 114 (1888): 113–53, 289–91.

way. However, if it was sloughed off, a different kind of abnormal development of half an embryo occurred (see figure 4.1). Roux concluded that each of the first two blastomeres developed independently of the other, and thus must do so under normal circumstances as well. The parts were not interacting with one another, and neither did it appear as if even the nucleus of the egg was interacting with the cytoplasm. Roux referred to this as self-differentiation, and it became known as **mosaic development**, in contrast to **regulative development**, where cell interactions and other influences exerted a regulatory control that determined final outcomes. He thought that, from the second cleavage on, development was a mosaic of at least four vertical independent pieces, each following a particular pathway. This research appeared to have vindicated His's idea that specific factors caused differentiation and that they were localized in particular parts of the egg.

Roux's experiments were groundbreaking, but his conclusions eventually had to be drastically altered. Development of half embryos did occur up to the neural stage, yet some of those half embryos gradually formed a complete whole embryo. The simplest way to interpret these results was that the determinants or factors must not have been

destroyed. All the factors still existed in each of the two cells after the
first cleavage. To save his idea, Roux suggested that there were two different kinds of determinants. One type was qualitatively divided and
specified the organs and other parts of the embryo. However, another
type was quantitative and was held in reserve. It would come into play
if a part was lost later in development, and it would enable normal generation to continue. Yet how could one even design an experiment to
test whether this was true? It seemed at best a kind of *ad hoc* explanation
to save his original hypothesis (see figure 4.2). Furthermore, in some
cases Roux found that whole embryos would develop from single cells
of the two-cell stage embryo. This argued against the idea that specific
factors were located in parts of the egg and were divided up.

Hans Driesch (1867–1941) continued Roux's line of research, trying
to definitively prove or disprove Roux's hypothesis of mosaic development. Instead of frog embryos, Driesch's organism of choice was the
sea urchin. Rather than killing one of the cells from the two-cell stage,
Driesch put about 100 embryos in a small tube with sea water and
shook it violently. When the cells of the two-cell stage stayed together,
the embryos exhibited mosaic development, just as Roux had observed.
Each developed into half of an older embryo. However, some of the
two-cell embryos had broken through their membrane and the cells
had separated. These separated cells each developed into normal
pluteus (larval stage that looks much different than the adult) larvae,
although smaller. Indeed, even up to the fourth cleavage, each cell was
able to develop normally into a complete individual. This contradicted
both Roux and His. It also suggested that perhaps there are not universal laws of development. Sea urchins are not frogs. These findings
instead indicated that normal frog development was the product of
independent self-differentiating parts, but that the whole embryo also
exhibited some kind of overall regulatory control. As Driesch wrote,
the sea urchin embryo was a "harmonious equipotential system."
Nevertheless, he became frustrated when he could not isolate what was
responsible for the regulatory control. Instead he postulated that an
immaterial force he called "**entelechy**" was what guided development.
He also hypothesized that perhaps Roux had not truly isolated blastomeres at the two-cell stage in the frog embryos. He had killed one
cell, but when it remained in contact with the living cell, it still exerted
an overall inhibitory regulative effect. Driesch was later shown to be

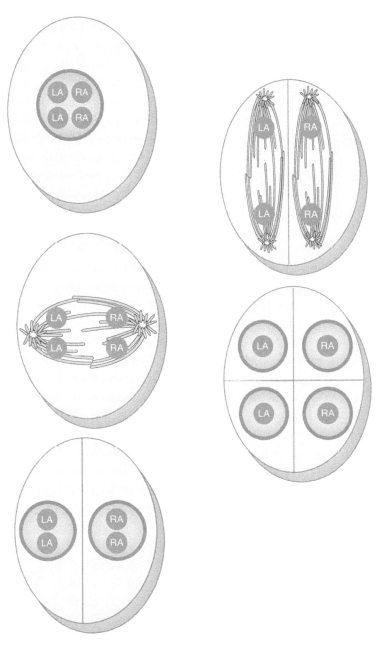

4.2 Schematic of Roux's revised hypothesis that becomes known as mosaic development. The segregation of determinants results in a mosaic of at least 4 vertical pieces developing independently. LA = left anterior determinants; RA = right anterior determinants; LP = left posterior determinants; RP = right posterior determinants.

correct. In 1910, J.F. McClendon was able to truly isolate one cell of the two-cell stage frog embryo by sucking it out with a tiny pipette, and a normal but small frog larva developed. Nevertheless, in spite of Driesch's excellent research suggesting that other cells and the whole organism shaped development, his advocacy of entelechy contributed to organicism being associated with vitalism and unclear thinking. He eventually abandoned biological research altogether for philosophy.[21]

These studies with sea urchin and frog embryos seemed to suggest that His's hypothesis was incorrect. However, as briefly mentioned in chapter 3, Morgan's studies with centophore embryos seemed to be a prime example of purely mosaic development. The bodies of these beautiful medusa-like invertebrates were transparent; like the sea urchin, this made them an ideal organism to study early development. They moved though the water propelled by eight rows of comb plates. Following the first three cleavages of the egg, Morgan got the following results. If blastomeres were isolated from one another in the two-cell stage, cleavage occurred as if each cell was still in contact with the other, developing as a half embryo. The larvae only had half the number of rows; however, each row had a full complement of the combs or swimming paddles. If the blastomere was isolated from the four-cell stage, it sometimes developed into a one-fourth larva with only two rows of paddles. If three of the four blastomeres were kept together, the embryo developed with six rows of paddles. Later experiments demonstrated that a blastomere isolated from the eight-cell stage would produce a larva that only had one row of comb plates. Such results were extremely confusing. What kind of mechanism allowed mosaic development at the two-cell stage with each cell developing independently of its respective parts, but allowed each isolated cell to still be capable of developing an entire organism? As in the case of the sea squirts, it was apparent that some kind of regulatory mechanism was in place that allowed the parts to form more than they normally would if they were isolated. These experiments supported the idea that the cell even within an organism was independent, and at the same time also provided evidence in support of Huxley's criticism of cell theory. In multicelled organisms the cell was not fully independent, but was being informed by other cells within the embryo as development unfolded.

---

21 Gilbert and Sarkar, "Embracing Complexity," 4.

Another advocate of mosaic development who contributed to its support was August Weismann (1834–1914). Most histories of biology have emphasized Weismann's contributions to genetics and evolutionary theory. In a famous experiment to disprove Lamarckian inheritance, he cut off the tails of 68 rats over five generations, but none of the 901 offspring were born without tails or altered ones. Weismann is most known for the somatic/germ line distinction. This was important for understanding the significance of meiosis. Not only does the production of four **gametes**, each with half the number of chromosomes, prevent the doubling of the hereditary material each generation, but Weismann also argued that the **germ plasm** destined to produce gametes were sequestered from the rest of the body. They were thus protected from any environmental influences. Since these cells were what contributed to the next generation, this effectively precluded Lamarckian inheritance. This was contrasted to Darwin's concept of **pangenesis**. Darwin thought that particles from the somatic cells, which could be influenced by the environment, could migrate to the germ cells and change them. However, what is important for our story at this stage is that Weismann, like others of his generation, considered heredity essential to understanding development – and the central problem was to understand how cell differentiation occurs as development proceeds. His soma/germ line distinction was a crucial aspect of his embryological theory. He and Roux agreed that the first cell contained all the determinants for development, but as cell division proceeded these determinants were divided unequally among the cells. Roux's embryological experiments seemed to suggest this as well. Yet this mosaic theory of development did not explain how new structures were actually formed. The cells that were destined to become muscle cells carried the determinant to become muscle, just as the skin cells carried the determinant to become skin, but the theory did not explain how this happened.

Although Weismann had made important contributions to the significance of meiosis, it was Oscar Hertwig who discovered and described meiosis for the first time in sea urchin eggs in 1876. He was highly critical of Roux's and Weismann's claim that development was mosaic. Although Hertwig thought that heredity was localized in the nucleus, he attacked the mosaic theory on several grounds. Just as Morgan had initially not wanted to locate the hereditary material in the nucleus because it smacked of preformation, Hertwig thought the same of these hypothetical determinants. What exactly were these determinants in the germ cells that could not be seen? An even more profound problem was that even if inheritance was particulate, the theory also assumed a causal relationship between the determinant and the body parts that developed. Hertwig instead argued that cell differentiation occurred as a result of the interaction with other cells. Nevertheless, for both the advocates of mosaic development, who claimed some kind of preformed particles existed, and those who claimed development was regulated by the organism as a whole, the fundamental problem remained: how were new parts made and where did they come from?

At the Zoological Station in Naples, Wilson was also trying to solve the puzzle of whether development was regulative or mosaic. He began his investigations on the amphioxus or lancelet, another primitive marine organism that was of interest to embryologists because it had characteristics of both vertebrates and invertebrates. Instead of a spinal column made of vertebra, it had a rod of connective tissue called a notochord. Using the same basic technique as Driesch, Wilson vigorously shook the dividing embryo until the individual cells fell apart. When isolated at the two-cell stage, development proceeded essentially normally, producing a blastula and gastrula that were normal but only half the usual size. Some continued to develop into normal but dwarf-sized larva up to when the first gill slits formed. However, development of blastulas isolated from the four-cell stage varied widely. Some began normally, but most did not, and those that did rarely developed as far as the larva stage when the notochord appeared. Those blastula isolated from the eight-cell stage also seemingly started to develop normally, but none achieved the gastrula stage. These findings suggested to Wilson another plan of research that would prove to be extremely useful: tracing the fate of each cell. Wilson hypothesized that every cell contained all the genetic material, but that the cells became modified as development progressed, as a result of their interactions with other cells of the embryo. However, they still retained all the hereditary elements. Isolating the blastomeres in the very early stages resulted in the cell returning essentially to the original state of the single egg, and thus it was able to develop normally. However, by the eight-cell stage the genetic material had become sufficiently modified that the cells were no longer capable of being, in the words of Driesch, "equipotential," or in our modern-day terminology "**totipotent**." It appeared that as development progressed, the cells lost this capability and become more and more fixed in their pattern of differentiation. As Wilson wrote, "The independent self determining power of the cell, therefore, steadily increases as the cleavage advances.... In the earlier stages the morphological value of the cell may be determined by its location. In later stages this is less strictly true and in the end the cell may become more or less completely independent of its location, its substance having fully and permanently changed."[22] Wilson was claiming that each developmental step was determined in part by the processes and interactions between the cells in the preceding step. Logically this meant that the entire developmental pathway, which obviously could be interfered with at any stage, was dependent on the initial organization of the undivided egg. However, it also meant that development was mosaic: "The only real unity is that of the entire organism, and as long as its cells remain in continuity they are to be regarded, not as morphological individuals, but specialized centers of action into which the living body resolves itself, and by means of which the physiological division of labor is effected."[23]

Europeans, particularly the Germans, had been the leaders in research concerning development. As we have seen, Americans such as Wilson and Morgan went abroad to study. However, by the end of the nineteenth century, the Americans developed their own experimental traditions.[24] One of their most important contributions was emphasizing the importance of **cell lineage** studies. From the uncleaved egg the products of cell division were traced, following each cell in development until the rudiments of the embryonic organs became distinct. In order to do this, one must be able to distinguish the individual cells from one another, either by their position, their size, or their color. Different species of marine invertebrates turned out to be the best source of suitable embryos. Centophores and sea urchins had transparent embryos, allowing one to see the cells in the interior. In many species, the dividing cells had distinctive patterns of coloration

---

22 Edmund Beecher Wilson, quoted in Moore, *Science as a Way of Knowing*, 456. For original, see Wilson, "Amphioxus and the Mosaic Theory of Development," 606–10.
23 Wilson, "The Mosaic Theory of Development," vol. 2, 9.
24 See Maienschein, *Transforming Traditions in American Developmental Biology*.

and/or when they divided gave rise to different-sized cells, making it easier to follow them. Also, as mentioned, the plane of the first cleavages was not random in relationship to pigmented areas, and in many cases the differently pigmented areas of the eggs seemed to have a specified relation to the germ layers and structures that would form. These findings suggested that the cytoplasm was highly organized. Development was not just determined by the hereditary material from the nucleus that was eventually shown to be located on the chromosomes. Nevertheless, experimental results did not unequivocally support this position. Impressed by some of Driesch's experiments in which two eggs fused and still formed a single embryo, some researchers claimed that the uncleaved egg had no axial organization, and that all parts of the cytoplasm were equivalent. Cell lineage studies could potentially resolve these two different ideas.

CELL LINEAGE STUDIES

The beginning of cell lineage studies is usually credited to Charles Otis Whitman, who in the 1870s described the cleavage patterns of the leech embryo. Like Wilson and Morgan, he also went to the Naples Zoological Station. In 1888 he became the founding director of the Marine Biological Laboratory in Woods Hole, Massachusetts. Woods Hole became a leading institution in studies on heredity and development, which continue to this day. Whitman extended his studies of the leech to many other organisms, including sea urchins, nematodes, and sea squirts. He observed that in some groups, such as the nematodes and sea squirts, the pattern of the cell division was virtually identical from individual to individual. This made it possible to construct detailed lineage trees following the fate of particular cells. However, in other groups, such as leeches and insects, the invariant patterns occurred as sublineages, that is, they were seen in the progeny of particular precursor cells. In both of these cases, because of the correlation between the lineage and the cell fate, Whitman and others assumed that the fate of a particular cell was determined by factors that were segregating within the dividing cells. This was called "determinate" cleavage. Whitman showed that in the leech, the fourth cleavage gave rise to four very small cells and four large ones.

The four small cells eventually gave rise to the **ectoderm**, while one of the four large cells that he called *x* gave rise to the **mesoderm** and the nervous system. He eventually demonstrated that complete organ systems could be traced back to a particular pair of cells. These findings, along with Roux's early views, supported His's hypothesis. Whitman wrote, "the future of the embryo was determined right from the beginning.... The egg is in a certain sense, a quarry out of which without waste, a complicated structure is to be built. It is the architect of its own destiny."[25]

For Whitman development was determined, not regulative, and is referred to as **determinant development**. In addition, Whitman's studies in the bony fishes caused him to think the early successive stages were not dependent on cell division. He wrote, "may we not go further, and say that an organism is an organism from the egg onwards, quite independently of the number of cells present."[26] Whitman thought the forces that determined when and why cells divided did not pay any attention to cell boundaries, and molded the germ mass, regardless of the way it was cut into individual cells. The division of cells was of secondary importance. For Whitman, comparative embryology demonstrated "the organism dominates the cell formation, using for the same purpose one, several, or many cells, massing its material and directing its movements, and shaping its organism, as if cells did not exist, or as if they existed only in complete subordination to its will."[27] As mentioned in the previous chapter, Whitman not only claimed that cells were of secondary importance, but that cell theory was actually interfering with achieving a full understanding of development. What he referred to as the *cell-standpoint* was the idea that in understanding cell formation lies the whole secret to development. Whitman disagreed with this idea. In "The Inadequacy of the Cell-Theory in Development," he analyzed why regarding the cell as the primary unit of structure, as an independent individual, was misguided. Describing the development of organism after organism, Whitman showed "that organization precedes cell-formation and regulates it, rather than the reverse, is a conclusion that forces itself upon us

---

25 Charles Otis Whitman, quoted in Moore, *Science as a Way of Knowing*, 458. For original, see Whitman, "The Embryology of Clepsine," 263–64.

26 Whitman, "The Inadequacy of the Cell-Theory in Development," 645–46.

27 Whitman, "The Inadequacy of the Cell-Theory in Development," 653.

from many sides."[28] He thought there were some unknown elements, that he gave the name *idiosomes*, that "were the bearers of heredity and the real builders of the organism. " A modern reader would probably conclude that Whitman's *idiosome* would turn out to be DNA. However, this would miss the true significance of his critique. First, by claiming that idiosomes were the bearers of the hereditary information, but also that they were responsible for growth, organization, and differentiation, Whitman again is arguing that development and heredity are but two aspects of the same fundamental problem – generation. They can't and shouldn't be separated. But for the purposes of this discussion, Whitman's take-home message was that what determined the action and control of these elements was *independent* of cell boundaries: "The organization of the egg is carried forward to the adult as an unbroken physiological unity, or individuality, through all modifications and transformations." If this sounds remarkably similar to Huxley, that is because it is. Whitman, in fact, ended his article with a quote from Huxley's review of cell theory about the relationship of shells to the tides (see chapter 2), claiming it "forms a fitting conclusion to this introductory sketch."[29]

Several other investigators working on different organisms at Woods Hole also provided evidence that development was determined right from the beginning. In their cell lineage studies, Wilson, working on the marine worm *Nevis*, and E.G. Conklin, working on the limpet *Crepidula*, discovered that the early cleavage pattern of these two organisms was virtually identical. Both formed a particular cell from which all the later mesodermal structures were derived. Such findings demonstrated that the developmental patterns were determined early on. In addition, since the organisms were from different phylums whose adult structures were very different from each other, these results provided powerful evidence for Darwin's theory of common descent. Conklin was also able to follow in splendid detail the developmental stages of the mature egg all the way to a fully formed larva in the ascidian (a type of tunicate) *Cynthia* (now *Styela*). This was because the egg had a large clear germinal vesicle. The inside consisted of a mass of gray yolk, while the periphery contained yellow pigment, but fertilization resulted in a

---

28 Whitman, "The Inadequacy of the Cell-Theory in Development," 650.
29 Whitman, "The Inadequacy of the Cell-Theory in Development," 658.

dramatic rearrangement of the cytoplasm. Conklin discovered that the colored regions corresponded to the boundaries of the germ layers, and thus he could trace the fate of the cells. The yolky cells gave rise to **endoderm**, the yellow crescent region formed muscles and mesenchyme, while the clear cytoplasm of the animal pole gave rise to all the ectodermal structures. He was even able to distinguish what area would form the **neural plate** and the **notochord**.

Frank R. Lillie (1870–1947) was another early pioneer of the American school of embryology. Born in Toronto, he came under the tutelage of Whitman at the University of Chicago and Woods Hole. He spent his entire career in various academic institutions in the United States, including establishing the Woods Hole Oceanographic Institution and serving as its first president in 1930. Early in his career, by subjecting eggs of an annelid worm to abnormally high concentrations of potassium, he showed that normal development could proceed quite far – even in the absence of mitosis or cell multiplication. Lillie interpreted these findings to mean that cell division was not the primary factor in embryonic differentiation. Rather, the primary role of cell division was to serve as a process of localization. Lillie also made much broader claims in regard to the relationship of the parts to the whole. The prevailing view was that initially development might be mosaic, the parts acting independently, but that physiological unity arose by secondary adaptation of one part to another. However, Lillie disagreed, claiming that this view was no longer tenable. Rather, there was a unity of organization that was part of the original inheritance, cycling through generation after generation. August Weismann agreed and claimed that what was responsible for this organization was the chromosomes located in the germ cells. (Were these Whitman's idiosomes?) As we already saw and will continue to discuss in the next chapter, the rapid progress made in the new field of genetics was going to provide a great deal of support to such a position. However, Lillie did not agree with Weismann's position either. Instead, he argued that the organization was not due to any substance within individual cells. Rather, "[t]he cells are subordinated to the organism which produces them and makes them large or small, of a slow or rapid rate of division, causes them to divide, now in this direction.... The organism is primary, not secondary; it is an individual not by virtue of the cooperation of countless lesser individuals, but as an individual that produces these lesser individuals.... The persistence of organization is a primary law of embryonic development."[30]

These findings of Whitman, Wilson, Lillie, and Conklin continued to provide more and more evidence that the egg was highly organized and that during the course of normal development particular areas of the egg determined the structures of the growing embryo – again just as His proposed many years earlier. Nevertheless, as informative and meticulous as these studies were, they were still primarily descriptive as to what happened in normal development. This did not mean that particular cells could *only* give rise to particular structures. One could not unequivocally conclude that structures from the older embryo could *only* come from specific precursor cells. As embryologists came to realize, one had to distinguish between the *fate* of the cell and the *capacity* of the cell. During the course of normal development a particular cell might be destined to form a particular structure, but it might still retain the capacity to form other structures under different developmental circumstances. The distinction between regulative and mosaic development not only applied to the whole embryo, but also applied to the developing parts. Once again, Huxley's assertion that the cell was not an independent unit suggested a research program that was worthy of further investigation: to unravel how the cell's fate and its capacity were determined by interactions with other cells and its environment. Whether development was mosaic or regulative, whether it was primarily determined by the nucleus or the cytoplasm, was still an open question. How cell differentiation occurred was essentially a mystery, and watching normal development was not going to solve it. The end of the nineteenth century saw the end of the "gentlemen naturalists" who primarily observed and described. The biologists of the twentieth century were "hard-nosed materialists who sought to understand the principles of living systems in order to bring them under human control."[31]

Ross Harrison (1879–1959) was another experimental embryologist whose research was heavily guided by cell theory, but at the same time embodied the organicist perspective in his quest to understand development. He developed the technique of tissue culture to investigate

---

30 Frank R. Lillie, quoted in Ritter, *The Unity of the Organism,* vol. 1, 14. For original see Lillie, "Observations and Experiments," 222.
31 Allen, "Mechanism, Vitalism and Organicism," 275.

the nature of nerve outgrowth. Profoundly interested in problems of symmetry and form, he investigated how the axis of symmetry was established in the developing limb bud and inner ear of the newt. He showed that the limb developed as a "harmonious equipotential system," which was shaped by gradients and the spacial organization of the cells to each other. His work contributed to the concept of **morphogenetic fields** in the 1930s. Structure and function were tightly linked. Like Hertwig, he recognized that "it is impossible to develop science wholly from the top down or from the bottom up."[32] Harrison was an outstanding example of someone who pursued a nonvitalist, experimental approach to development within an organismic framework.

In a series of remarkable, detailed, and delicate experiments, the Swedish embryologist Sven Hörstaudius (1898–1996) separated and recombined in different ways the various layers of developing sea urchin embryos and followed the fate of particular cells. From these experiments, he hypothesized that the egg and early embryo contained particular animalizing and vegetalizing substances that followed a concentration gradient. The animalizing substances were necessary for the structures that normally developed from ectodermal areas, while the vegetalizing substance was necessary for the structures that formed from the mesoderm and endoderm. If this was the case, normal development might not be determined by the particular parts involved, but rather the proper balance of these two hypothetical substances. His results provided evidence for such an idea. Normal development occurred when the ratio between the two substances was approximately equal, intermediate when there was more of an imbalance, and extremely abnormal if the ratio was markedly unequal. If blastomeres were isolated from the two-cell stage, and the separation was along the animal-vegetable axis (A-V), then each half embryo would have the entire range of these hypothetical substances and normal development occurred. However, if the first cut was equatorial, separating the animal and vegetable poles, then development was abnormal. Experiments that followed the development of the isolated halves indicated that development was regulative when separation was along the meridian or A-V axis, but was mosaic if the separation occurred

---

32 Gilbert and Sarkar, "Embracing Complexity," 4.

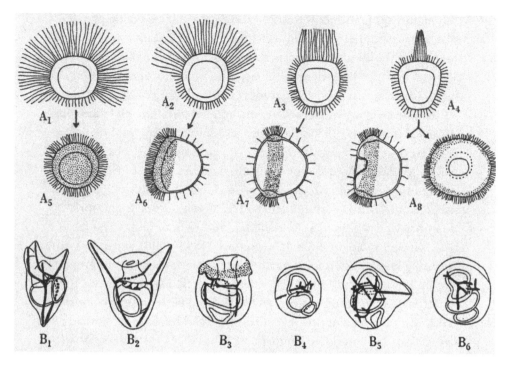

4.3 The development of isolated animal (A) and vegetal (B) hemispheres of sea urchin embryos (Hörstadius, 1939). Reproduced from S. Hörstadius, "The Mechanics of Sea Urchin Development, Studied by Operative Methods," *Biological Review* 14 (1939): 132–79. Copyright © by John Wiley & Sons, Inc. or related companies. All rights reserved.

along the equatorial plane. Finally, although it appeared that normal development depended on concentration gradients of particular substances along the A-V axis, the substances were not localized in particular areas. As development proceeded, one could eliminate any one of the particular tiers of cells and a normal larva would still develop. Thus, development of a part depended on the entire embryo (see figure 4.3). Cells were not independent. Furthermore, the development of a part was regulated in such a manner that the end result would be as normal as that particular isolated fragment would permit. These experiments showed that one could not definitively describe sea urchin embryonic development as either regulative or mosaic. It depended on the condition and the parts being discussed. Driesch's conclusions from his early experiments were not wrong, but they were incomplete.

Cell lineage studies also irrefutably showed that the egg was highly organized, and as division proceeded certain cells became destined to form particular structures. At the same time they also revealed how complex the whole process was. Development could be considered to be either mosaic or regulative, depending on the organism and also what stage of embryogenesis one was observing. These studies again confirmed Huxley's caution about the limitations of cell theory in trying to understand development. They showed that a continual complex interaction was occurring between cells to ensure normal development. Furthermore, this indicated that the whole organism had to be considered the primary unit of organization that was overseeing the entire developmental process. The role of the individual cell was of secondary importance. However, for the most part, this research did not translate into biologists explicitly acknowledging the problematic aspects of cell theory. One exception was William Emerson Ritter (1856–1944).

WILLIAM RITTER AND
THE UNITY OF THE ORGANISM

William Ritter's story is both important and interesting, as it illustrates the scientific and extra-scientific considerations that shaped twentieth-century biological investigations. As the history of science shows us time and time again, many factors influence the acceptance or rejection of specific ideas at a particular moment in history. Ritter was highly critical of cell theory, particularly in relationship to understanding how life developed. His disapproval of cell theory was part of a much larger agenda, in which he wanted to reformulate and change the whole theoretical basis of biological research.[33] However, he became increasingly marginalized in spite of doing much outstanding work. In most histories of biology he is barely mentioned, if at all.

Ritter was born in Wisconsin and started his career as a public school teacher. However, very interested in learning more about the biological world, he moved to California, attended Berkeley as an undergraduate, and then continued his studies at Harvard, obtaining his PhD in 1893.

---

33 See Esposito, "More than the Parts," 273–302.

Like so many others in our story, he also went to the Naples Zoological Station where he studied tunicates. He eventually returned to the States and became a zoology instructor at Berkeley. Ritter not only shared Huxley's concerns about cell theory, but like Huxley he also became a tireless advocate for science education. They both were committed in their belief that furthering the understanding of the natural world and promoting scientific thinking would help alleviate human misery. Ritter conducted his research on marine organisms, but for different reasons than most of the biologists in our story.

As we have seen, research was becoming more experimental and interventionist. Rather than mere description, one had to bring organisms back into the laboratory, examine them under the microscope, dissect them, and perform various manipulations. However, Ritter went against this increasingly prevailing view, believing that it was more important to study organisms in their natural environment. In keeping with his interest in science education, he wanted to teach marine biology by taking students to the ocean where they would learn about organisms in their native habitat. With encouragement from others and help from the newspaper magnate E.W. Scripps and Scripps' half-sister Ellen, he helped found the Marine Biological Association of San Diego and became its scientific director. This eventually became the Scripps Institution of Oceanography of the University of California, San Diego. Today, it remains true to Ritter's vision. A leader in research, its mission statement is "to seek, teach, and communicate scientific understanding of the oceans, atmosphere, Earth, and other planets for the benefit of society and the environment."

Ritter had this larger goal in relationship to science education, but it was driven by his quest to understand what made something "alive." It can't be overemphasized, and therefore is worth repeating, that until the turn of the twentieth century heredity and development were treated as complementary aspects of the fundamental problem of generation. Ritter never accepted the split that had occurred between the two fields and maintained such an approach would never lead to a full understanding of the organism. However, Ritter was not just advocating an old view. He recognized the distinction between the two disciplines, but he did not agree with the theoretical underpinnings of the research agenda that was separating them. Instead, he pursued investigations that unfortunately resulted in his becoming isolated from scientists in both disciplines. In what he considered his magnum opus, *The Unity of the Organism* (1919), Ritter offered a critique of what he called the elementalist conception of the organism.

A key part of Ritter's analysis was a close examination of cell theory and why he thought it was not sufficient to explain the living organism. Cell theory had played a critical role in finally putting to rest the debate over spontaneous generation. However, some people who accepted cell theory still advocated what could be considered a form of vitalism. Vitalists argued that there was some kind of vital or spiritual force that animated life. Huxley argued that there was no essential difference between a rock and a fish – it was just a matter of chemistry. As Huxley had written, even thought could be explained in terms of chemical reactions. However, Huxley also would have agreed with Ritter, who advocated organicism. As mentioned in the beginning of the chapter, organicism predates Ritter and it was the basis for Hertwig's materialistic and interactive theory of development. Whitman also promoted what he referred to as the "organismal standpoint." It is Ritter, however, who is credited with being the first person who used it to build a biological theory that claimed that life resulted from the interrelationship of living beings. For Ritter, life could not just be reduced to physics and chemistry. Even molecules exhibited properties that could not be predicted simply by looking at the individual properties of the atoms that made them up. This was true of organisms as well. As Ritter argued, "life ... is the sum total of the phenomena exhibited by myriads of natural objects called living because they present these phenomena." Furthermore, in agreement with Huxley and in the tradition of Cuvier, to truly understand the function of the parts, they must be seen within the context of the whole organism.

Ritter maintained that both historically and logically the "*organism is made to do duty in interpreting the cell.*"[34] For Ritter, the organism was prior and contributed to the idea of the cell. The two existed in a reciprocal relationship and, as he pointed out, it can be thought of as the old problem of the chicken and the egg. Is the chicken the way the egg makes another egg, or is the egg the way the chicken makes another

---

34 Ritter, *The Unity of the Organism*, vol. 1, 156; emphasis in original.

chicken? By 1919 Weismann's ideas, that the germ cells were sequestered off and contained the hereditary material, were fully accepted.
This finding had given dominance to the idea that the germ cells interpret the parent. Ritter thought this was, however, only a half truth. The
parents also interpret the germ cells. The two were causally related,
each following the other, so neither one was prior. This had been most
dramatically illustrated in plants. The form of the growing tip often
occurred before the mass actually divided into the cells. Thus the whole
plant formed cells; the cells did not form the plant. For Ritter, like
Whitman, this meant that the formation of cells, while essential, was
still a secondary phenomenon: "They should be regarded as *organs* of
the organism, just as muscles and glands and hearts and eyes and feet
are so regarded."[35] Furthermore, he argued that thinking of the cell
as an elementary organism assumes that the organismal perspective is
more fundamental. Just as the components in the cell are organized,
forming a single organism, cells are likewise parts of the integrated
organization of the whole organism.

Morgan still had a physiological view of development that was independent of cell boundaries, but he did not share Ritter's views. When
Ritter wrote to Morgan in 1911, reminding him of his earlier embryological studies that contradicted his genetic findings, we find no evidence that Morgan ever replied. As Morgan continued his genetic
studies, he moved ever farther from his initial views concerning development. Whereas he had earlier claimed that cells were interacting and
communicating with each other and their environment, and this was
what was critical to understanding development, later he increasingly
emphasized the importance of the nucleus. By 1934 his position on
morphallaxis was totally the opposite of what he had initially thought.
The cells reoriented themselves by the differential activity of the genes.
Nevertheless, one could still have asked what was responsible for the
differential action of the genes, but this was a problem that at the time
was not solvable, especially since scientists still did not actually know
what a gene was.

One would have thought that Wilson would have been a great
supporter of Ritter in light of his many earlier statements about the
importance of the whole organism. Yet when Ritter raised issues

---

35 Ritter, *The Unity of the Organism,* vol. 1, 158, 191; emphasis in original.

concerning the limitations of cell theory and objecting to the separation of heredity from development, it appears that Wilson, like
Morgan, did not reply. Wilson continued his work in development
and accepted the chromosomal theory of inheritance even before
Morgan did. Since genetics became essentially the study of the transmission of the heredity units, which did not influence Wilson's own
cell lineage work in any significant way, he did not find the split problematic. Furthermore, his views seemed to be changing about the role
that the whole organism played in shaping development. In earlier
writings he claimed that "the life of the multicellular organism is to be
conceived as a whole." Yet in his 1900 edition of *The Cell* he wrote, "the
key to all ultimate biological problems must, in the last analysis, be
sought in the cell."[36] Ritter thought this view was entirely misguided.
In addition, Ritter thought that the debate over whether development
was predetermined or epigenetic had confused the importance of the
whole organism in development. There was a unity or oneness of the
organism though time and through space. Quoting Huxley, "the germ
is not merely a body in which life is dormant or potential, but that it
is itself simply a detached portion of the substance of a pre-existing
living body."[37]

Nevertheless, even if one shared Ritter's holistic views, it was difficult to design experiments with such a model. Many, if not most,
researchers recognized that the organism was more than the sum of
its parts, with emergent qualities arising that could not be predicted.
However, in order to get a handle on generation and the complexity
of development, it was necessary to begin somewhere. As Medawar
claimed, research is the art of the soluble. The separation of heredity and development into two distinct disciplines turned out to be
particularly powerful in understanding the laws of inheritance and
identifying the hereditary material. Ritter's position became increasingly ignored, as the research agendas of genetics and embryology
became quite distinct with virtually no overlap. There would be no
serious attempt to bring heredity and development back together for
several decades. Ritter's story is significant as it highlights not only
how cell theory in certain ways impeded a more organicist/holistic

---

36 Wilson, quoted in Ritter, *The Unity of the Organism*, vol. 1, 162.
37 Huxley, quoted in Ritter, *The Unity of the Organism*, vol. 1, 165.

approach to development, but also that Ritter recognized heredity and development would need to be reunited to untie the Gordian knot of development.[38]

The next chapter examines how this split resulted in spectacular advances in our understanding of heredity, but doing so also meant that the fundamental question that Huxley had posed 100 years earlier – "How does form come to be?" – did not receive much attention from the vast majority of researchers. That question will be revisited in chapter 6.

---

38 See Esposito, *Romantic Biology 1890–1945*.

CHAPTER FIVE

Progress in Understanding Heredity

Genetics is about how information is stored and transmitted between generations.

John Maynard Smith, 1999

Our story so far shows that before the early twentieth century heredity and development were intimately linked. The advent of cell theory along with improvements in microscopy brought attention to the nucleus. But its function was still unknown, and a debate ensued over whether the nucleus or the cytoplasm directed development. As evidence mounted that the hereditary material was located in the nucleus, heredity was still not considered to be distinct from development. As Edwin Conklin wrote, "Indeed, heredity is not a peculiar or unique principle for it is only similarity of growth and differentiation in successive generations.... The causes of heredity are thus reduced to the causes of successive differentiation of development and the mechanism of heredity is merely the mechanism of differentiation."[1] However, this would soon change, and studies in heredity and development became quite separate from each other with the emergence of the new discipline of genetics.

---

1 Conklin, "The Mechanism of Heredity," 90.

The outstanding success of genetic investigations in the first half of the twentieth century culminated with the elucidation of the structure of DNA and how it carried out its function as the hereditary material. This resulted in genetics providing the dominant approach for understanding a variety of biological problems from evolution to cancer, and continues to this day. However, as we have seen, part of that spectacular achievement was also due to redefining what heredity meant. Heredity came to mean transmission genetics. Rather than being an essential component of embryological development, heredity came to be regarded as totally independent of development. It is worth examining how this division came about in a little more detail as it had enormous ramifications for how cell theory was regarded in the first part of the twentieth century. In addition, the split shaped the research agendas of not only genetics, but developmental biology and evolutionary theory as well.

NARROWING THE MEANING OF HEREDITY

It is surprising that Thomas Morgan, who began his career as an embryologist, was one of the people most responsible for the split that resulted in developmental and hereditary studies becoming quite separate in the twentieth century. Before his work on *Drosophila* chromosomes convinced him of the importance of chromosomal genes, he had agreed with Conklin, writing, "We have to look at the problem of heredity as identical with the problem of development."[2] Furthermore, he had been skeptical of any particulate theory of inheritance. Mendelism, the chromosome theory of inheritance, and Darwin's theory of pangenesis were all particulate theories and, therefore, they had preformationist implications. Since development of the embryo was the result of processes leading to increasing complexity, it could not be due to a series of particles that were responsible for adult traits such as blue eyes or orange fur.

However, as previously pointed out, Morgan reversed his position on the role of chromosomes as a result of studying the developmental causes of sex determination. Discovering mutations that segregated with

---

2 Morgan, "Chromosomes and Heredity," 449.

the sex chromosome in *Drosophila* convinced him that the Mendelian laws correlated with a chromosomal theory of inheritance and that both must be correct. As these ideas gained more acceptance, however, simplifying assumptions were also made. As Mendel's factors metamorphosed into genes, the idea of the gene and the unit character it was associated with often became conflated. Morgan, true to his embryological training, still insisted that they were not the same. He recognized that some factors or genes affected many characters (later defined as **pleiotropy**), and that many genes virtually always affected a single character. Indeed, he identified 25 different loci on the chromosome that affected the red eye color in *Drosophila*. When one of those factors mutated, the result was a pink eye. Although all the other factors still affected the eye color as well, Morgan and his co-authors wrote that the mutated factor was the cause of the pink eye color: "[W]e use cause here in the sense in which science always uses this expression, to mean that a particular system differs from another system only in one special factor."[3]

Defining the gene as the cause of a particular character was, however, a truly radical claim and has been glossed over in most histories of heredity. Prior to this, the cause of any adult trait could theoretically include the entire developmental history of the organism. Reducing the cause of pink eye color to a single gene effectively eliminated any discussion of the developmental processes in the explanation of how a gene actually is responsible for the development of the pink eye color.[4] Furthermore, the authors knew exactly what they were asserting. They acknowledged that genes did not explain embryonic cell differentiation, but also that they weren't trying to assert that. Instead, they argued, "it stands as a scientific explanation of heredity because it fulfills all of the requirements of any casual explanation."[5] Morgan put the final nail in the coffin of the earlier meaning of heredity, which had heredity and development inextricably interconnected, in his 1926 book *The Theory of the Gene*. He argued that a lot of the criticism of genetics was the result of confusing problems of genetics that were actually questions concerning development. This was highly ironic, since many of the criticisms had, in fact, been his own! But now he insisted that gene theory

---

3 Morgan et al., *The Mechanism of Mendelian Heredity*, 209.
4 Amundson, *The Changing Role of the Embryo*, 148–58.
5 Morgan et al., *The Mechanism of Mendelian Heredity*, 226–27.

could be explained without any reference to trying to connect the causal sequence between the gene and its character. Genetics (the term was coined by William Bateson in 1906) now referred to how genes sort themselves from generation to generation. This was accomplished by an analysis of the character traits they correlated with, but totally left out any reference to the causal connections between the two.

Morgan did make a distinction between two types of genetics. The first was transmission genetics, which meant Mendelian genetics, and the second was developmental genetics, which was the study of the physiological action of genes in embryonic development. He suggested that embryologists should devote themselves to this kind of study. The end result was that transmission genetics took ownership of the word "heredity."[6] While one of the underlying themes of this book is that heredity and development must be reunited to truly understand how a cell becomes an organism, the split did result in stunning progress in our understanding of this new discipline of genetics. Nevertheless, while many characters do follow Mendelian laws of inheritance, many do not. Furthermore, the implication that all hereditary similarities were due to Mendelian genes remained controversial for many years. Moreover, defining heredity as the transmission of Mendelian factors was an assertion that made a distinction that many embryologists objected to. They also did not accept that studying the physiological processes of genes would fully unravel the problem of development. It would be several decades before significant numbers of the biological research community would realize the embryologists were basically correct. Nevertheless, it is also worth looking at what this split allowed the new field of genetics to achieve.

THE RISE OF POPULATION GENETICS

Redefining heredity narrowly to mean only transmission genetics removed the difficult problem of development from the discussion, but the study of **mutations** and their effects received criticism from other quarters as well, not just from embryologists. Mendel's laws applied to discrete characters, but did not seem to explain the continuous

---

6 See Gilbert, "Bearing Crosses," 168–82.

variation that naturalists were observing in the field. Traits such as the color, height, weight, and shape of a plant were not discrete. Instead, they exhibited a bell-shaped distribution within certain limits. This was a major difficulty for advocates of Darwin's theory. Darwin had argued that natural selection acted on very small variations, producing change slowly and gradually. However, there was a problem: could this mechanism actually create novel traits? Natural selection might be able to adapt an organism to its environment, but did it have the power to push the population beyond the limits of variation observed in the parent population and actually create new species? Interestingly, this was something that Huxley had argued about with Darwin, claiming that saltative evolution (evolution by jumps) better explained the pattern of the fossil record, and that true speciation had not yet been observed either in nature or by artificial selection. Huxley eventually changed his mind about saltation and the fossil record, but he remained, in Darwin's words, his "Objector General" in regard to natural selection.[7] Bateson agreed with Huxley, but also recognized, as Darwin had claimed, that variation was the raw material of evolution. Thus Bateson thought the best way to study evolution was to study variation. Working in the field, he identified several species that had distinct forms with no intermediates. He concluded that speciation was the product of discontinuous variation. He started breeding experiments to understand the laws of inheritance of these discrete discontinuous variations. Rather than natural selection being the primary mechanism of evolution, the idea that speciation was due to the appearance of large variations was reinforced by several researchers, particularly Hugo de Vries. Although de Vries had independently rediscovered Mendel's laws, he did not think they provided any additional insight into how species actually arose. He also began intensive breeding experiments with the evening primrose (*Oenothera*) and cultivated two plants that were quite distinct from the normal type found in the wild. They bred true, but also threw out new forms that also continued to breed true. They were so distinct that he felt justified in calling them new species. Several new forms appeared in a single step. De Vries called these new discontinuous variations "mutations," which later came to be seen as the same kinds of mutations that Morgan was breeding in his flies. These mutations were inherited while

---

7 See Lyons, *Thomas Henry Huxley,* 1999, chapters 5 and 7.

the other small variations due to environmental fluctuations were not. It was not until much later that it was realized that the evening primrose consisted of unusual hybrid species that were **polyploid**, and the mutations were not, in fact, what de Vries thought they were.

Mendel was analyzing his peas when Darwin was developing his theory, but Darwin was not aware of Mendel's work. Darwin's theory seemingly would have been on much stronger ground if he had understood the source of the variation for natural selection to act on. In yet another irony of history, Bateson, de Vries, and other Mendelians all used the rediscovery of Mendel's work to argue against Darwin's mechanism of natural selection. They accepted common descent and that organisms changed through time, but claimed that mutation, rather than natural selection, was the primary mechanism of evolution. Bateson introduced several terms to clarify the laws that Mendel and de Vries had discovered. "Segregation" referred to the separation of the hereditary units in meiosis and was responsible for the definite ratios that Mendel had observed. The units that segregated he called "**alleles**." Alleles that were the same when recombined in the zygote – for example, both alleles coded for round seeds – were referred to as being **homozygous**. If they differed from one another, one round and one wrinkled, he called this **heterozygous**. Yet naturalists working in the field observed that most traits did not segregate in this clear manner. Instead, they saw small variations and organisms well adapted to their particular environment. Because of this, naturalists thought new species arising by saltation due to large mutations seemed extremely unlikely. Nevertheless, the **mutation theory** enjoyed considerable popularity at the beginning of the twentieth century. The key to resolving these seemingly contradictory views was to unravel the genetics of continuous variation.

Biometricians Karl Pearson (1857–1936) and W.R. Weldon (1860–1906) pioneered a statistical approach to analyzing these small variations. The Danish biologist Wilhelm Johannsen (1857–1927) established several distinct pure lines of the garden bean that tended to self-fertilize and thus were largely homozygous. By careful statistical analysis, Johannsen showed that each line of beans varied continuously in size over a certain range. The range of variation that occurred in the offspring was virtually identical from generation to generation and remained constant regardless of what size bean he grew the plants from. As a result of this work, he introduced several new terms, using the term "gene" for the pangene of de Vries, and making the fundamental distinction between genotype and phenotype. The genotype referred to all the genes in the fertilized zygote. The phenotype was what an organism looked like as a result of the interaction between genes and the environment. The environment was responsible for the considerable variation that he observed in his pure lines. Herman Nilsson-Ehle (1873–1949) confirmed Johannsen's ideas, examining color inheritance in wheat grains. Red always was dominant to white, but varied in intensity. By crossing reds with whites, he found the offspring fell into three distinct groups, but each group exhibited a range of variation around a mean. He explained his results by postulating that three different genes contributed to the red color and that they segregated independently of one another. When crossing the homozygous white with a red, one red gene would be enough to turn the grain red, but the intensity of the red would be dependent on the number of red alleles that were inherited. Effects of the environment also influenced the growth and color and yielded offspring that showed a range of variation. The effects of the environment would further smooth out the differences, yielding a continuous variation of color, just like Johannsen had observed in his pure lines. Both men's work showed that only the variation due to the genotype, rather than the environment, was inherited. Most important, they showed that the individual mutation could be much smaller than de Vries envisioned. Instead, the continuous range of variation observed in the wild could be explained by multigene inheritance following Mendel's laws. Nevertheless, Johannsen thought that natural selection acting on continuous variation could not shift the genotypes. It could only prefer one to another within the population. Thus, mutation was still what was responsible for creating something new for selection to act on. The mutations might be small, but mutation was the driving force of evolution. In Morgan's fly room, a continual stream of spontaneous mutations were appearing that he used for his breeding experiments and that provided evidence for such a position. Natural selection was still not well accepted.

Building on the work of biometricians and the breeding experiments of Nilsson-Ehle, William Castle (1867–1962), and others, a new group of geneticists pioneered a mathematical approach that combined biometrics and Mendelian genetics. They looked at the effects of selection and systems of mating to build quantitative models of the evolutionary process. The new field of population genetics was born. In

1908 G.H. Hardy (1877–1947) showed what would be the distribution of a pair of alleles A and a in a population with no selection. The possible genotypes were aa, AA, or Aa. If the alleles were present initially in a ratio of A = p and a = q, then p + q = 1. Using Mendel's laws of inheritance, and assuming independent mating, the frequency of the alleles in the next generation can be determined by the equation $p^1 + 2pq + q^1 = 1$. This showed that the proportion of the two alleles A and a would remain constant in future generations if nothing upset this equilibrium. Independently Wilhelm Weinberg (1862–1937) came up with the same idea, which is now known as the Hardy-Weinberg principle. However, this was a highly simplified model of what happened in nature and was virtually never true. Selection, nonrandom mating, mutation, migration in and out of the population – all were factors that would disturb this equilibrium. Nevertheless, the mathematician H.T.J. Norton (1886–1937) was one of the first people to recognize the significance of Hardy's work, as he showed that even a slight advantage of 10 percent or less could change the frequency of a gene in a population. Selection *was* powerful. Ronald Fisher (1892–1962) continued to build on these ideas, modeling the effects of variation, drift, mutation, and selection on gene frequency. He demonstrated that evolution occurred gradually by selection acting on small genetic differences, just as Darwin had argued. Fisher's book *The Genetical Theory of Natural Selection* (1930) summarized the key ideas that had emerged. First, there was no inherent tendency for variability to diminish over time. Second, genes lost through drift or by chance had significant effects only in small isolated populations. Third, since mutations were relatively rare and most tended to be deleterious, they would be quickly weeded out by selection. Fourth and most important, Fisher showed by his calculations that selection was by far the most important factor in changing the gene frequencies of a population. J.B.S. Haldane (1892–1964), another mathematical biologist, came to the same conclusions and applied his calculations to a real case: industrial melanism of the peppered moth. The directional shift of the color of the moth population resulted from air pollution, and the dark-colored variant was favored as that color made the moths less visible to predators. When pollution declined, the light-colored variant came to predominate again, as it blended in well with the white lichens that covered many of the trees. Castle also provided additional evidence of the power of selection and the importance of gene interactions in his breeding experiments with different colored hooded rats. The black-and-white hooded pattern appeared to be inherited as a simple recessive to the dominant gray. Yet the pattern was quite variable, and by selective breeding Castle was able to establish quite stable strains that were practically all white or all black. To return them to the original pattern was as difficult as the original selection. Back-crossing the strains with homozygous gray rats resulted in offspring with highly variable amounts of black and white in their coat color. This showed that a trait was not determined by only a single gene, but instead could be modified by the interaction with other genes. This provided further evidence for the power of selection and the idea that a single gene influenced a variety of different traits. Multiple effects of single genes in *Drosophila*, which resulted in each offspring being a little different from the parents, had been well documented. These various breeding experiments in different organisms showed that there was an enormous store of variation "hidden" in the genome available for natural selection to act on. As Castle wrote, this meant that natural selection was an agency of real creative power, capable of far more than just weeding out deleterious traits or selecting favorable ones.

Castle's student Sewall Wright (1889–1988) continued Castle's breeding experiments with rats, and also did his own on the inheritance of coat color in guinea pigs, which he continued long after he left Castle's lab. He developed new methods to analyze the results and came to the conclusion that selection was most effective on groups of interacting genes. As a result of constant inbreeding, he found that the families became more and more different from each other, but also more and more homozygous. It appeared that certain genes became lost while others became fixed, but it was a matter of chance as to what combinations became fixed in particular families. He applied his findings to evolution in the wild and argued that selection would be most effective in small populations, which by chance might have frequencies of certain genes that were not representative of the species as a whole. This in turn would lead to combinations that became fixed, unlikely to occur in a larger population because of gene flow among individuals. Selection would be able to act more efficiently on these combinations in the small populations, and the species would evolve more quickly. This became known as Wright's "shifting balance theory of evolution."

That genetic drift and selection in small populations played a more significant role than selection acting on single genes in large populations caused quite a dispute between Wright and Fisher and Haldane. Wright later changed his mind about the importance of drift, and by the 1960s relegated it to a relatively minor role in evolution. In spite of the differences between Wright and the other two men, they all agreed that mutation by itself was relatively powerless to cause a significant change in gene frequencies. This went against the ideas of Bateson and de Vries. Furthermore, what had appeared to be two different kinds of variation, the continuous variation found in nature and the discontinuous variation observed in artificial selection, were shown to follow the same rules of Mendelian inheritance. The variation observed in nature was the product not only of the action of the environment, but also of the fact that many genes influenced the same trait. Due to multiple-gene inheritance, it was possible to produce populations that exhibited variation that went beyond the limits of that observed in the parent populations. Population genetics modeling had demonstrated unequivocally that over time new species could result from selection acting gradually on discrete and usually quite small mutations. This was a critical finding, since one of the most consistent criticisms of Darwin's theory from Huxley to de Vries had been that natural selection did not have the power to actually create new species, and that mutation was the primary factor driving evolution.

Population genetics paved the way for an era of collaborative research in the 1940s that brought the ideas of Darwinians and Mendelian geneticists together, giving rise to what Julian Huxley (the grandson of Thomas) dubbed the Modern Synthesis for evolutionary theory. While many people's research contributed to the synthesis, they credit more than anyone else the Russian population geneticist Theodosius Dobzhansky.[8] His 1937 book *Genetics and the Origin of Species* brought together the research of naturalists working in the field and geneticists in the labs, melding their findings into a coherent theory. The Modern Synthesis vindicated Darwin's claim that evolution occurred by the gradual accumulation of very small inherited variations, and that

---

8 Several important books were published during this period, including Huxley's *Evolution*, Simpson's *Tempo and Mode in Evolution*, Mayr's *Systematics and the Origin of Species*, and Stebbins's *Variation and Evolution in Plants*, as well as Dobzhansky's book.

the primary mechanism of change was natural selection. Mendelism triumphed, and Lamarckian forms of inheritance along with mutation theory were thoroughly discredited. Evolutionary theory joined cell theory in being one of the few great unifying ideas for biology. As Dobzhansky would later write, "Nothing in biology makes sense except in the light of evolution."[9]

The synthesis period, however, might more accurately be referred to, in the words of historian Will Provine, as the "evolutionary constriction," in that genetics came to totally dominate the theoretical aspects of moving the discipline forward. Paleontological investigations and field research were labeled as merely descriptive, and relegated to the role of providing confirmation of the abstract formulations of the increasingly sophisticated modeling of the geneticists.[10] Botany also got short shrift in the synthesis period, in spite of the fact that Darwin had made extensive use of botanical data from breeding experiments and the geographic distribution of plants. Certain groups of plants exhibited a saltative pattern of evolution that was revealed to be due to large chromosomal rearrangements, as was the case of the evening primrose. Most significantly for our story, as Huxley had pointed out in his critique of cell theory, plants in particular were exceptions to the idea that cells were independent units. These areas of research, along with embryology, were essentially left out of the synthesis, also in spite of the fact that Darwin had regarded embryology as the most important evidence in favor of his theory of common descent. Thus, separating embryology and heredity into distinct disciplines had important consequences for both evolutionary theorizing and understanding development.

WHAT ARE GENES?

The narrow redefining of the meaning of heredity also meant that well into the twentieth century geneticists seemed quite uninterested in the physical nature of genes. It was enough to know that they were located on the chromosome for scientists to do their research. Indeed,

---

9 Dobzhansky, "Nothing in Biology Makes Sense," 125.
10 See Provine, "Progress in Evolution and Meaning of Life."

in his 1934 Nobel laureate lecture in Stockholm, provocative as always, Morgan made what seems an astonishing claim to a modern reader. He argued that it didn't matter whether genes

> are real or purely fictitious – because at the level at which the genetic experiments lie it does not make the slightest differ-ence whether the gene is a hypothetical unit or whether the gene is a material particle. In either case the unit is associated with a particular chromosome, and can be localized there by purely genetic analysis. Hence if the gene is a material unit, it is a piece of chromosome; if it is a fictitious unit, it must be referred to a definite location in a chromosome – the same place as in the other hypothesis.[11]

The geneticists may not have been interested in the physical character-istics of heredity, and as we saw in previous chapters, it took some time to establish that the nucleus played a significant role in directing devel-opment. It took even longer to identify that chromosomes were being passed on generation to generation. Morgan may have suggested that genes might be fictitious, but chromosomes were not. They were real. However, they also disappeared; they could only be seen when cells were actively dividing. Yet the hereditary material had been discovered more than 50 years earlier.

In 1869 Friedrich Miescher (1844–95) isolated from the pus of sol-diers a substance that he called **nuclein**. This substance contained DNA (deoxyribonucleic acid) and its associated proteins that came from the cell nuclei. He eventually separated out DNA as a distinct mole-cule. Nevertheless, just as Hooke's conception of a cell had very little to do with our modern conception of the cell, likewise Miescher did not recognize the importance of his discovery. Rather, he thought that the molecules of heredity would be protein – an idea that would be extremely hard to displace for a variety of reasons. Miescher knew it was critical to find the chemical foundation that was responsible for the differential staining between the nucleus and the cytoplasm, but he did not succeed in this task. Others had claimed to find what was responsi-ble for the specificities observed, but when he couldn't duplicate it in

---

11 Morgan, "The Relation of Genetics to Physiology and Medicine," 315.

his own work, he dismissively referred to the cytologists as dyers. How wrong Miescher was about the role cytology played and continues to play in understanding life processes! Unraveling the workings of the cell needed a multipronged attack. Building on Miescher's work, biochemists continued to separate and characterize the constituents of the nucleus. E.B. Wilson was one of the first people to suggest a chromosomal model for heredity and that the chromatin might be the physical correlates of inheritance. As he wrote in 1895,

> These facts justify the conclusion that the nuclei of the two germ-cells are in a morphological sense precisely equivalent, and that they lend strong support to Hertwig's identification of the nucleus as the bearer of hereditary qualities. The precise equivalence of the chromosomes contributed by the two sexes is a physical correlative of the fact that the two sexes play, on the whole, equal parts in hereditary transmission, and it seems to show that the chromosomal substance, the chromatin, is to be regarded as the physical basis of inheritance. Now, chromatin is known to be closely similar to, if not identical with, a substance known as nuclein which analysis shows to be a tolerably definite chemical composed of nucleic acid (a complex organic acid rich in phosphorus) and albumin. And thus we reach the remarkable conclusion that inheritance may, perhaps, be affected by the physical transmission of a particular chemical compound from parent to offspring.[12]

As Wilson pointed out, chromatin consists of both **nucleic acid** and **protein**. It would take a long time before scientists gave credit to nucleic acid as the carrier of the hereditary information.

The German biochemist Albrecht Kossel (1853–1927) succeeded in isolating from the nucleus five organic compounds that would later be shown to be the bases that made up the structure of DNA and **RNA**. Kossel was interested in doing more than just strict chemical analysis. He thought chemistry should become less statistical and more dynamic in order to understand the role that organic compounds played in

---

12 Wilson and Leaming, *An Atlas of the Fertilization and Karyokinesis.*

metabolic pathways. From his physiological studies, he concluded that nucleic acids were neither storage nor energy compounds. They were confined to the cell nucleus and, furthermore, they were not found throughout the nucleus, but rather to what histologists called chromatin. Recognizing the importance of this discovery, he wrote, "This fact is of major significance for the relationship of chemistry to the cell theory, for it gives us a chemical characterization for an elementary organ of the cell in addition to the morphological characterization." The chromatin was a mixture of both nucleic acids and protein, and Kossel thought that nucleic acids represented unique building blocks that could be the foundation for investigating the chemical activity of the nucleus. A modern reader might think that surely Kossel would conclude that the function of nucleic acids was the transmission of genetic information. Yet he did not. Instead he wrote that its activity "is to be sought in relation to growth and to the replacement of protoplasm."[13] Kossel's ideas typified the thinking of biochemists up into the 1950s. They were interested in metabolic pathways, the synthesizing and degradation of molecules, which had its parallel in what was happening to the cell as a whole. The emerging group of researchers who would call themselves molecular biologists, however, were interested in the problem of cell replication. Nevertheless, Kossel made many important contributions. Perhaps most significant was that he originated the idea of the Bausteine or building block for cell chemistry. Metabolism was to be understood as the linking together and separating of basic building blocks, which in turn resulted in the synthesis and degradation of secondary complex molecules. As embryonic development proceeded, more and more Bausteines linked together, giving rise to varied complex proteins. However, nucleic acids did not seem to have nearly the variety or complexity of proteins, consisting of only five basic building blocks. Proteins, however, contained up to 21 different kinds of amino acids that could be linked together in any number of ways. Nucleic acids, with their seemingly simple structure, could not possibly be up to the task of carrying hereditary information.

Phoebus Levene (1863–1940) continued to build on Kossel's work and found that DNA consisted of approximately equal amounts of the four bases – **adenine**, **thymine**, **guanine**, and **cytosine** – along with the

13 Albrecht Kossel, quoted in Olby, *The Path to the Double Helix*, 75.

sugar deoxyribose and a phosphate group.[14] He also showed that each base was linked together with the sugar and phosphate to form a unit that he called a nucleotide. He suggested that the DNA molecule consisted of a string of nucleotide units linked together through the phosphate groups that formed the "backbone." He also thought that each DNA molecule contained only four nucleotides. This became known as the tetranucleotide hypothesis. Further work showed that DNA was a very large molecule, consisting of hundreds if not thousands of nucleotides linked together. Because of the approximately equal amounts of each base, it was thought that DNA consisted of repeating identical tetranucleotides. Once again, how could such a molecule have the biological specificity necessary to be the carrier of the hereditary information? Its structure was much too uniform to be able to generate complex genetic variation.

Biochemists were very slow to recognize the function of nucleic acids. However, they had led the way in showing the importance of macromolecules and developing the idea of individuality and biological specificity. Molecules needed to be thought of as more than just substances in metabolic pathways. Cytochemists also played a critical role because they had shown that the chromosome was made of nucleic acids and protein. Since proteins existed in diverse forms while DNA was regarded as a "stupid molecule," they also perpetuated the idea that genes would be proteins. What emerged was the nucleoprotein theory of the gene, which would shape the research for several decades. As the geneticist/cytologist Cyril Darlington wrote, "Cell genetics led us to investigate cell mechanics. Cell mechanics now compels us to infer the structure underlying it. In seeking the mechanisms of heredity and variation we are thus discovering the molecular basis of growth and reproduction. The theory of the cell revealed the unity of living processes; the study of the cell is beginning to reveal their physical foundation."[15] Once again microscopy would be essential to the cell giving up its secrets. Yet cytologists and biochemists clung to the idea that it was the protein part of the nucleoprotein that was important. This led them to increasingly unlikely interpretations of their data,

---

14 It would later be shown that ribonucleic acid (RNA) contained uracil instead of thymine along with the other three bases.

15 Darlington, "Interaction Between Cell Nucleus and Cytoplasm," 932.

and it would be many years before the true significance of DNA was recognized.

Improved staining techniques revealed that the chromosome had a definite and constant morphology and was made up of specific segments that had characteristic banding patterns. It was now possible to do a qualitative analysis of the chromosome. This allowed geneticists to construct more and more detailed gene maps by studying the chromosomal rearrangements of known genetic characters. This meant that trying to sort and analyze the respective roles nucleic acids and protein played seemed both possible and necessary to fully understand how chromosomes carried out their hereditary function. Boveri had established that the individuality of the chromosomes was maintained through successful resting stages, and he suggested that the reason they "disappeared" was because they lost the substance that was responsible for taking up the stain in the chromatin. At the onset of division they became "recharged." When it was shown that chromatin was a mixture of protein and nucleic acid, but it was the nucleic acid that was taking up the stain, this suggested that the protein component was what persisted. This was interpreted to mean that protein must be the critical part of the chromosome. As Wilson wrote in his very influential *The Cell in Development and Heredity*, "The nucleic acid component comes and goes in different phases of cell activity."[16] At this time relatively little was known about the detailed stages of the cell cycle in terms of synthesis and segregation of specific molecules. The only phases in which chromosomes could even be observed was in mitosis and meiosis.

Miescher's "dyers" continued to experiment with different kinds of reagents and an improved microtome, and also examined their preparations with the new Zeiss ultraviolet microscope. Torbjorn Caspersson, Einar Hammarsten, and Harold Hammarsten examined their chromosomal preparations in both living and nonliving cells. Much to their and everyone else's surprise, they found that the characteristic banding patterns that made possible chromosome mapping was due to nucleic acid. Yet instead of inferring that the nucleic acid component was the important part of the nucleoprotein theory of the gene, they concluded just the opposite! "There are only proteins to be considered because

16 Edmund Beecher Wilson, quoted in Olby, *The Path to the Double Helix*, 103. For original, see Wilson, *The Cell in Development and Inheritance*, 653.

they are the only known substances which are specific for the individual. On that assumption the protein structure of the chromosomes takes on a very great interest."[17] The prevailing view was still that only proteins were capable of individual specificity. The nucleic acid essentially provided the scaffolding, maintaining the structural integrity of the chromosome, and holding the proteins in a particular position that resulted in the specific banding pattern. When the authors discovered that vigorous nucleic synthesis occurred at the same time the chromosomes were duplicating, they had to go through an even more convoluted interpretation in order to save the protein version of the nucleoprotein theory of the gene.

As more and more experiments continued to show the importance of nucleic acid, a slightly different version of the nucleoprotein theory emerged. Both protein and nucleic acid were important, and biological specificity was not due solely to protein. However, neither was it due only to nucleic acid. Both were essential to the individuality of the gene, but it was also thought the continuity of the chromosome was due to protein fibers. Thus, even in this version it was still thought that the storage of genetic information resided in the proteins, since that was what was passed on generation to generation. The nucleic acid was, however, needed for the duplication and expression of it. In the chromosomes of higher organisms, the nucleic acid was bound to histones and also associated with other proteins in a very complex manner. Various attempts to remove either the nucleic acid or protein part of the chromosome to show that it did not fall apart were not particularly enlightening, giving contradictory results.

In 1939, as a result of their enzymatic studies on the salivary chromosome, Daniel Mazia and Lucena Jaeger asserted that the structural integrity of the chromosome was due to different proteins and that nucleic acids were not necessary. The chromosomes were completely digested by trypsin, an enzyme that hydrolyzes protein. If treated with a nuclease, the chromosome no longer stained with techniques that would indicate the presence of nucleic acid. However, it still absorbed stain, indicating the presence of protein. These results showed that the structural

---

17 Torbjorn Caspersson, quoted in Olby, *The Path to the Double Helix*, 105. For original, see Caspersson, Hammarsten, and Hammarsten, "Interactions of Proteins and Nucleic Acids," 369.

framework of the chromosome was not affected by nucleic acid. But if protein held the chromosome together *and* supposedly carried the genetic information, then what exactly was the function of nucleic acid? The synthesis of nucleic acid corresponded with the duplication of the chromosome, but the integrity of the chromosomes was due to the proteins. With hindsight it is hard to understand chemists' insistence that proteins rather than nucleic acids must be the hereditary material. This demonstrates how biases and preconceived ideas influence how one interprets experiments. Many years later, Mazia wrote that chemists always lagged behind biologists in imagining and accepting new ways of thinking about the cell and its functions.[18] This was written from the perspective of a cell biologist and one who in his own career would continually suggest innovative ideas about the nature of cell processes. Whether one accepts Mazia's views or not, chemists and chemical cytologists had failed to identify the chemical nature of the hereditary material by both direct and indirect approaches. Something else was needed.

The study of chromosomes in the cells of higher organisms had not yielded their secrets. What about in **bacteria** and viruses? In 1928 Fredrick Griffith had identified something he called a "transforming principle" in pneumococcal bacteria that could transform one type to another. If dead bacteria from a virulent strain that produced smooth-looking colonies was injected into a mouse along with live bacteria from a nonvirulent strain that produced rough-looking colonies, the mouse would develop a fatal infection. Furthermore, live bacteria of the virulent strain could be isolated from the mouse. Somehow the nonvirulent strain had been transformed into a virulent strain. Previously, it was thought that only live bacteria of the virulent strain could cause infection. These were truly astounding results, but for the next decade most geneticists continued to ignore the findings in microorganisms, just as microbiologists ignored findings in genetics. Biologists working on higher organisms didn't even necessarily accept that bacteria were fundamentally the same as cells found in multicellular organisms. Bacteria had no visible chromosomes. Neither microbiologists nor cell biologists thought that the genetics of microorganisms could be anything like those of higher organisms.

---

18 Daniel Mazia, June 1983, personal communication.

Griffith's work was quickly confirmed, and many bacteriologists became interested in trying to find the nature of this transforming principle. They became convinced that it must be chemical. The appearance of the smooth surface of colonies in the virulent strain was due to the presence of a polysaccharide capsule, and thus it seemed to be a reasonable candidate as the transforming substance. This turned out not to be the case. In 1944 Oswald Avery, Colin MacLeod, and Maclyn McCarty purified a substance that could transform the rough cells into smooth cells, and by a variety of tests they identified it as DNA.[19] Treatment with various enzymes that destroy protein left the transforming principle intact. Ribonuclease, which denatures RNA, also did not destroy the activity. However, treatment with deoxyribonuclease destroyed its activity. Was DNA in fact the gene? As Dobzhansky had suggested earlier in regard to Griffith's work, "If this transformation is described as a genetic mutation – and it is difficult to avoid so describing it – we are dealing with authentic cases of inductions of specific mutations by specific treatments – a feat which geneticists have vainly tried to accomplish in higher organisms."[20]

Since Mazia had shown that the structural continuity of the chromosome was due to protein, not nucleic acid, he initially thought the genetic material must be protein. However, as more and more experiments were indicating that DNA was the hereditary material, Mazia pointed out that if so, it must fulfill certain expectations. As one of the foremost researchers on mitosis, he emphasized the properties relating to the cell cycle. First, the hereditary substance should be quantitatively the same in every **diploid** cell of a species and should also be quantitatively proportional in the **haploid** cell. Second, the material should exactly double somewhere in the mitotic cycle. Third, it must also be very stable, although he recognized it was difficult to set an exact criterion of how it should be stable. Fourth, it has to be capable of specificity, that is, capable of having a large variety of individual configurations. Finally, it must be capable of being transferred from one cell to another and obtaining the same results as when genes were introduced by genetic techniques.[21]

19 Avery, Macleod, and McCarty, "Studies on the Chemical Nature," 137–58.
20 Dobzhansky, *Genetics and the Origin of the Species*, 49.
21 Daniel Mazia, quoted in Olby, *The Path to the Double Helix*. For original, see Mazia, "Physiology of the Cell Nucleus," 111.

The tide had finally begun to turn. Trying to save the protein version of the nucleoprotein theory of the gene was becoming increasingly difficult with more and more experiments pointing to nucleic acid as the active component. In addition to the Avery-MacLeod-McCarty experiment in 1944, a crucial experiment – famously known as the Waring blender experiment – showed that the transforming activity was due to nucleic acid, not protein. The Waring blender had been invented in the 1930s by Fred Osius, who went to the popular band leader Fred Waring for financial backing. Who knew that an appliance for mixing cocktails was going to play a decisive role in displacing the protein version of the nucleoprotein view of the gene?! In 1952 Alfred Hershey and Martha Chase demonstrated that the transforming principle was indeed DNA.[22] Viruses are essentially nucleic acids encased in a protein coat. Phages are viruses that infect bacteria and, by differentially radioactive labeling either the protein or DNA component, Hershey and Chase were able to identify which component was active. Phosphorus (P) is found in DNA but not protein, while sulfur (S) is found in protein and not DNA. The phages were grown in media containing either radioactive phosphate or radioactive sulfur isotopes before infecting the bacteria. Making use of the bacteria's cell machinery to reproduce, the resulting phage progeny would contain the radioactive isotopes in their structures. The experiment was performed once using ^{31}P and once with ^{34}S. Unlabeled bacteria were then infected by the labeled phage progeny, which left the phage coat outside the bacterial cell while the genetic material entered. Agitation in a Waring blender followed by high-speed centrifugation separated the phage coat from the bacteria. The bacteria were then lysed to identify the new phage progeny. The progeny from phages that had originally been labeled with radioactive phosphate remained labeled while those labeled with radioactive sulfur did not. Further experiments provided evidence that the activity was not the result of small amounts of protein contamination, although some people were still convinced that DNA could not possibly be a complex enough molecule to be the genetic material. Even Hershey and Chase did not claim that DNA was solely the hereditary material. Rather, they concluded that protein was unlikely to be it and that DNA

---

22 Hershey and Chase, "Independent Functions of Viral Protein and Nucleic Acid,"
    39–56.

had some unspecified function. Biases are not given up easily! It would be physicists, x-ray crystallographers, and physical chemists who illuminated the structure that unequivocally identified DNA as the genetic material.

THE RISE OF DNA

DNA you know is Midas' Gold. Everyone who touches it goes mad.

Maurice Wilkins, 1979[23]

A small group of scientists thought that finding the correct structure of DNA would be crucial for understanding how the molecule worked. They included the renowned American chemist Linus Pauling (1901–94), the brilliant British physicist Francis Crick (1916–2004), who because of serving during World War II had still not finished his PhD at age 34, and the brash young American biologist James Watson (1928–). Watson and Crick teamed up in search of the structure at the Cavendish laboratory in Cambridge, England. Also working on the problem was the chemist and x-ray crystallographer Rosalind Franklin (1920–58), and physicists R.G. Gosling (1926–2015) and Maurice Wilkins (1916–2004), at Kings College London. In 1953 Watson and Crick proposed a structure of DNA followed by a confirmatory paper by Franklin and Gosling, which ushered in a new era of biological research that can only be described as revolutionary. It was not just that they discovered the structure, but it was their approach that was novel, as they made use of data from a variety of different disciplines. From Pauling they took the idea of model building. Pauling, with his profound understanding of the possible configurations of various atoms, had shown that some proteins were helical by making use of atomic models. The biochemist Erwin Chargaff had disproven the tetranucleotide hypothesis, but had shown instead that the amount of thymine (T) always equaled the amount of adenine (A), and the amount of guanine (G) always equalled the amount of cytosine (C). However, the ratio of AT to GC varied, depending on the organism the DNA was isolated from. Most

---

23 Maurice Wilkins, quoted in Judson, *The Eighth Day of Creation*, 70.

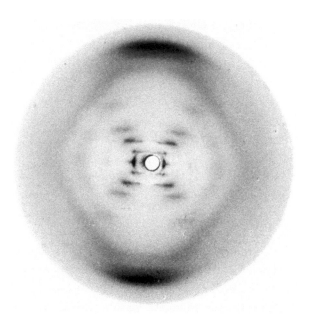

5.1. X-ray diffraction photograph of the B form of DNA. Photo by Rosalind Franklin. Image from College Archive collection, King's College London.

important, Watson and Crick made use of an x-ray diffraction image, taken by Franklin, that became known as photograph 51 of a crystalline form of DNA, which was absolutely essential to building their model (see figure 5.1).

Watson and Crick began their paper with what could be considered one of the greatest understatements in the history of biology, claiming that the structure "has novel features that are of considerable biological interest."[24] Building on what Mazia had suggested years earlier, the hereditary material must (1) have a way of duplicating itself, (2) have a way of encoding information, and (3) be able to exert a highly specific influence on the cell. In the 1950s and 1960s, scientists worked out in exquisite detail how DNA carried out all of these functions. The structure of DNA was confirmed, and details of DNA replication were explained. The discovery of several different classes of ribonucleic acid

---

24 Watson and Crick, "Molecular Structure of Nucleic Acids," 737.

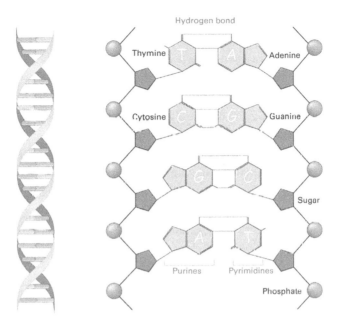

5.2. Schematic of DNA. Designua/Shutterstock.com.

(RNA) and the subsequent discovery of the "genetic code" through which DNA specified the different amino acids in a protein sequence enabled scientists to show how DNA's third function was carried out.

DNA is a long double-chained molecule, the two chains twisting around one another forming a double helix. Each chain consists of the sugar deoxyribose alternating with phosphate as its backbone. Attached to the sugar is one of four bases: the purines, adenine (A) and guanine (G), and the pyrimidines, thymine (T) and cytosine (C). The bases can be attached in any order along one chain, and the two chains are held together by hydrogen bonding between the bases (see figure 5.2). However, the base paring is very specific. A can only pair with T, and G only pairs with C. Thus A always = T and G = C, but the ratio of AT to GC can and does vary from organism to organism, just as Chargaff's data indicated. As Watson and Crick pointed out, "It has not escaped our notice that the specific pairing we have postulated immediately suggests a possible copying mechanism for the genetic material." The specific base paring of the two chains meant that each chain is complementary to the other. During replication the two chains could break apart, each serving as a template to make a new molecule. Two new molecules result, each made up of one old chain and one newly synthesized one. However, to prove that this was actually how replication occurred was not easy. In 1958 Matthew Meselson and Franklin Stahl demonstrated, in what has been described as "the most beautiful experiment in biology,"[25] that replication was semi-conservative, just as had been postulated. But what about the other two functions: encoding information and exerting a highly specific influence on the cell? While DNA had displaced protein as the bearer of hereditary information, proteins were still absolutely crucial to the functioning of the cell. Proteins not only make up important structural components of an organism such as muscle, fur, and blood, but they are also enzymes that drive virtually all the chemical processes that occur inside the cell. Thus, it was reasonable to think that DNA would contain information on how to synthesize proteins.

Someone who was extremely excited by the structure that Watson and Crick put forth was the Russian emigre George Gamow (1904–68), an astrophysicist, cartoonist, science popularizer, and at the time also a military strategist. He looked at the relationship between the structure of DNA and protein synthesis as a mathematic crypto-analytic problem. Gamow recognized that since the long sequence of bases can be in any order, virtually an infinite number of permutations were possible, and that could be the basis of a "genetic code" to specify each of the 20 amino acids that make up proteins. Since there were only four bases, Gamow quickly saw that a triplet code would be necessary; the code could obviously not be one base = one amino acid. Two bases would only yield 4^1 or 16 amino acids, but three bases would yield 4^2 or 64 possible combinations – more than enough to specify all the amino acids with plenty to spare. But how to find what triplet corresponded to what amino acid? Many schemes were tried, initially using DNA as the basis for the code. Crick, with characteristic bluntness, later said of their early ideas on protein synthesis, "Of course you realize that our ideas on that were totally wrong."[26] DNA is sequestered in the nucleus, and attention soon focused on the closely related molecule ribonucleic acid, named messenger RNA (mRNA). Researchers had hypothesized that it acted as

---

25 John Cairns to Horace Judson, in Judson, *The Eighth Day of Creation*, 188.
26 Francis Crick to Horace Judson, in Judson, *The Eighth Day of Creation*, 263.

a messenger, carrying the instructions from the DNA out of the nucleus into the cytoplasm for making protein. RNA was single-stranded, and instead of deoxyribose, the sugar is ribose, and **uracil** replaces thymine. Gamow, with his unbridled enthusiasm, founded the RNA Tie Club, whose aim was "to solve the riddle of RNA structure and to understand the way it builds proteins." This was to be a serious club, complete with charter. Gamow designed a tie that would be the club's emblem, brought to a "suitable haberdasher" to execute. The RNA tie was black and embroidered with brightly colored silk threads that represented the component parts of the molecule. The sugar-phosphate chain was in green and the bases were yellow. The club eventually had 20 members (one for each amino acid) and four honorary members. Each member was to also wear a tie pin that contained the three-letter abbreviation for his assigned amino acid. Gamow assigned himself alanine or ala and started addressing his fellow club members by what was on their tie pin, for example calling Crick Tyr for tyrosine. The club generated a lot of speculative ideas, and several papers got written. It is hard to convey the sense of excitement of these times, which have been called the "golden age of molecular biology," with so many innovative ideas and experiments taking place. The code was broken. Several **codons** (the sequence of three bases) coded for the same amino acid as well as a stop and a start codon were found. In a dazzling series of experiments, it was shown how the "stupid" molecule not only carried the hereditary information from generation to generation, but how it could direct so many activities of the cell.

This work also illustrates the deep connection between theory and experiment, as well as highlighting Crick's brilliance as a theoretician. This is not to downplay the importance of the significant contributions of many other scientists. However, Crick's ideas in particular point out how theory guides experiment. At the same time, theory ultimately has to be confirmed by experiment. Crick said that when building a theory, one should base it on as few facts as possible, which at first reading is quite counterintuitive and seems even ridiculous. Yet, as he pointed out, when you are truly at the frontiers of new knowledge, you don't know the difference between a right fact and a wrong fact. The whole history of science has demonstrated time and time again that what was believed as "fact" eventually turned out to be incorrect. For thousands of years it was accepted as fact that the sun went around the earth.

In the early eighteenth century combustible materials were thought to contain the substance phlogiston, which disappeared upon burning. Until Darwin, the prevailing view was that species were fixed and could not transform into other species. The list goes on and on. As Crick was working out the details of protein synthesis, he relied primarily on what was considered to be well known about the chemistry of the different molecules. As a result, he predicted the presence of an adapter molecule, when there was no evidence for it! The basic scheme the scientists were working from was that mRNA made a template from the DNA in the nucleus. This RNA carried the instructions for making a protein into the cytoplasm to the ribosomes, which were essentially protein factories. The ribosomes contained ribosomal RNA (rRNA), which enabled the mRNA to attach to the ribosome. However, Crick recognized that amino acids were not going to directly bind to the mRNA. The chemistry simply doesn't work. He hypothesized that there had to be another kind of adapter molecule that would be able to bring the amino acids to the mRNA. Based on Crick's idea, the search began. Transfer (tRNA) or soluble RNAs are small molecules that were very hard to find, but researchers found them. There is a specific tRNA for each amino acid. Further experiments showed that the amino acids do not directly bind to the mRNA, just as Crick suspected. Rather, the tRNA binds to mRNA, bringing each of the amino acids into place; they then are joined together.

The RNA tie club was dominated by physicists, most of whom had a certain disdain for the methods of the biochemists. Crick had decried the sloppiness of many of the biochemists' experiments, and also dismissed whole categories, including "the necessity for the rigorous experimental proof of each codon."[27] This is ironic on two counts. First, as just mentioned, it was those biochemical experiments that provided the necessary knowledge to allow him to predict the necessary existence of a molecule like tRNA. Second, it was Marshall Nirenberg (1927–2010), a biochemist and not a member the tie club, who unequivocally identified the first triplet code by a technique that allowed the breaking of the code to rapidly progress. Various researchers including Nirenberg had been experimenting with cell-free systems that used RNAs of mixed bases to make polypeptides. An RNA that contained

---

27 Francis Crick, quoted in Judson, *The Eighth Day of Creation*, 487.

only Us and Cs would incorporate the amino acid serine. The problem was that the bases were random, and so only their overall proportions could be obtained. Thus the experiment could not distinguish between UUC, UCU, and CUU. In 1964 at the sixth International Congress of Biochemistry, Nirenberg announced a major discovery. He and his postdoc Philip Leder had developed a technique that allowed them to make artificial lengths of RNA that were only three bases long. Yet it was long enough to make the ribosomes bind with the correct tRNA to carry the specific amino acids by that one-word message. They labeled one amino acid at a time and put the incubated mixture through a filter that held back ribosomes with the triplets and any transfer RNA bound to them, but let unbound tRNAS pass through.

DNA has often been called a beautiful molecule because its structure immediately suggested how it could carry out two crucial functions of the hereditary material: replication and carrying genetic information. The same cannot be said for most proteins. We now know the exact amino acid sequence of thousands of proteins, but knowing that has rarely provided any insight into how an individual protein carries out its particular function. There is another reason DNA is often described as a beautiful molecule. In addition to its structure suggesting how it carries out its hereditary function, several aspects of its structure are significant for evolution. First, the "code" is highly redundant. There are 64 different codons, but for the most part only 20 different amino acids are used in cell metabolism. Therefore, several codons code for the same amino acid. Usually, changing the third base of each codon results in the same amino acid being coded for. This is a safeguard that allows for a certain amount of inaccuracy in copying without creating any problems. An error in base pairing is much less serious than either an addition or deletion of a base because, if the latter occurs, the whole reading frame is shifted. Instead of one amino acid being off, potentially every one would be changed. Second, the hydrogen bonding between the two chains is very specific, but also relatively weak. Not only does that mean it can be broken easily for copying or transcribing, but it also means that base pairing errors will occur. We don't want too many, but errors are the raw material for natural selection to act on. Once the structure of DNA was known, mutations could be defined precisely. Mutations are changes in the sequence of bases in the DNA and are what create variability. Sometimes just a single change can make a profound difference. In sickle cell anemia, only one amino acid substitution has occurred, as the result of one base substitution. However, this is enough to create a defective hemoglobin molecule that causes the red blood cell to have a sickle shape. Yet in locations where malaria is prevalent, having one copy of this gene is beneficial, because the sickle cell is a less favorable environment for the malaria parasite to live in. This was one of our earliest examples of documenting natural selection at the molecular level. Many variants of the hemoglobin molecule have been found, and they may be beneficial in particular environments. Third, the genetic code is universal. It is virtually the same code in bacteria as in a human. This is why it is possible to insert the gene for human insulin into a bacterium and have it make human insulin. The bacteria can "read" the instructions encoded in the human gene correctly because they are both using the same universal language. Many different bacteria and other organisms such as yeast (important for the brewing industry) have been converted into molecular factories to make all kinds of molecules from virtually any species. The universality of the genetic code is powerful evidence that all life has descended from one or a few common ancestors.

Many people claim we are now in the second golden age of molecular biology. The discovery of DNA's structure ultimately led to the human genome project and has ushered in a whole new era of molecular medicine, with the potential for gene therapy and genetic engineering of organisms. It provided evidence for evolution at the deepest level, allowing us to investigate evolutionary processes at the molecular level. We now have sequenced the DNA of thousands of species, including ones that have gone extinct. The latest technology for modifying and editing DNA, CRISPR (Clustered Regularly Interspaced Short Palindromic Repeat), has changed the way basic research is being conducted. We now potentially have the means to alter the course of evolution, which has profound ethical implications. It also continues to fuel the emphasis of looking at the primary cause of disease as genetic, which in turn suggests that is where the bulk of research dollars on developing treatments should be as well.

No one can deny the spectacular success of the research program of molecular biology, particularly in our understanding of genetics. The underlying approach has been to try to reduce biological phenomena to the behavior of individual molecules. In spite of the truly impressive achievements of such an approach, discrete biological functions can rarely be attributed to the function of a single molecule in the sense that, for example, the function of hemoglobin is to transport gas molecules in the cell. Likewise, disease processes are complex and rarely come down to simply an error in the DNA. As mentioned, DNA has often been described as an elegant and beautiful molecule because its structure immediately suggested how it could carry out its function as the hereditary material. In addition, it was often called a self-replicating molecule. However, this is not the case. Unlike a crystal, DNA cannot replicate itself. Rather, it needs an enormous complex of enzymes as well as the organization of the cell to replicate. It accomplishes its biological function within the context of a host of other molecules as well as particular structures in the cell. Working out the details of how DNA carried out its function was suited to this highly reductive approach. In contrast, the origin of forms and their functions has been resistant to this approach, in part because it is intrinsically a much more difficult problem. The next chapter explores the research of embryologists who pursued a very different kind of research program. They maintained that to fully understand development, one could not work only at the level of molecules and cells.

CHAPTER SIX

Organicism, Embryonic Induction, and Morphogenetic Fields

Omnis organisatio ex organisatione.
Paul Weiss, *"The Problem of Cell Individuality in Development,"* 1940

As detailed in the last chapter, genetics made spectacular progress throughout the twentieth century, and in many ways overshadowed the work that was being done in embryology. However, significant progress was also being made in solving the enigma of development, as embryologists tried to understand the process of cell differentiation. Several biologists recognized the necessity of having a more holistic approach in their experimental design, including Hans Spemann (1869–1941), Paul Weiss (1898–1989), Joseph Needham (1900–95), and C.H. Waddington (1900–75). Building on the ideas of Driesch and Roux, rather than focusing at the level of the cell, they organized their research program at the level of tissues and organs. In doing so they all provided evidence for many of Huxley's objections to cell theory.

The last chapter illustrated that the history of heredity progressed in a more or less linear fashion. Yes, there were many blind alleys, and certainly the bias in favor of proteins probably delayed somewhat the discovery that DNA was the hereditary material. Nevertheless, progress in genetics can be characterized primarily as a series of successive, cumulative discoveries. However, as earlier chapters have shown, progress in understanding development occurred in a much more piecemeal fashion. As we will see in this chapter, various tensions continued to pervade the different research programs. What role did the whole organism play in directing development? How did the parts of the developing embryo influence each other? Researchers were also acknowledging that the role of genes could no longer be ignored.

THE WHOLE AND THE PARTS

German-born Hans Spemann was one of the greatest experimental embryologists of the twentieth century and was the first embryologist to win a Nobel prize for his work on induction and the concept of the **organizer**. He continued to build on the research of Hörstaudius. In contrast to Loeb, who worked at the level of the cell and thought that a complete explanation of development would be found at the level of molecules, Spemann's research focused on understanding how one part of the embryo influenced the other parts. A master of microsurgical technique, Spemann performed a series of novel and delicate experiments. He worked on a variety of organisms, but his most exhaustive experiments were conducted on amphibian embryos. He tied tiny hairs around the embryos, pulling on them very carefully until they were constricted into a dumbbell shape. Constricting the blastomeres of a fertilized salamander egg resulted in a partial double embryo with two heads and one tail. In older embryos, very different outcomes resulted, depending on where the constriction occurred. If the constriction crossed the blastopore (the slit-like invagination of the gastrula through which cells move to form internal organs), two complete embryos were created. However, if he tied the hairs above or below the blastopore, the region containing the blastopore developed into a complete embryo, and the region without it formed a soon-to-die undifferentiated belly mass. Spemann concluded that the embryo's blastopore region was essential for differentiation. However, by late **gastrulation** it was not possible to get duplicate heads or tails by constriction. Some kind of differentiation had occurred that limited the ability of the parts to develop normally.

Spemann and his graduate students continued to do groundbreaking **transplantation** experiments that provided evidence for what eventually became known as **embryonic induction**: that the development of a particular structure or tissue was affected by the neighboring tissue. He first demonstrated this in his studies of the vertebrate lens in frog embryos. Embryonic eyes start developing from optical vesicles coming from mesodermal tissue. Later, upon contact with ectodermal tissue, the ectoderm invaginates, then forms an optic cup and eventually the lens. Spemann found that he could transplant this ectodermal tissue far from where the eyes would normally develop and still induce the formation of a lens. In the converse experiment, he could also replace local ectodermal tissue with ectoderm from another part of the embryo and the lens would still form. Spemann concluded that the ectoderm tissue was predisposed to form eyes. Further work found that different tissues were predisposed to form particular structures, but were dependent on neighboring tissue to induce that development. In a series of detailed experiments, he and his student Hilda Proescholdt Mangold showed that mesoderm tissue from the front of the dorsal lip of the blastopore was necessary to cause the formation of the neural tube from ectoderm. The tissue that induced the formation of a particular structure became known as the "organizer," and where that structure formed was referred to as the organization center. The clear demonstration of the organizer concept showed that the cells were given instructions by neighboring cells (see figure 6.1). As Spemann (1943) remarked, "We are standing and walking with parts of our body which could have been used for thinking had they developed in another part of the embryo."[1]

Spemann's concept of development was very similar to Driesch's, who viewed the embryo as an **equipotent** harmonious system, but unlike Driesch he was not a vitalist. Because Driesch did not think that development would ever be explained by simple mechanical or reductive approaches, he felt the necessity to invoke his somewhat mystical concept of entelechy. Spemann did not agree. He demonstrated it was possible to design experiments that studied interactions at higher levels of organization and that allowed for causal analysis. This generated concrete hypotheses that could be tested. Throughout the first part of the twentieth century, research focused on finding more and more

---

1 Hans Spemann, 1943, quoted in Gilbert and Bard, "Formalizing Theories of Development," 131.

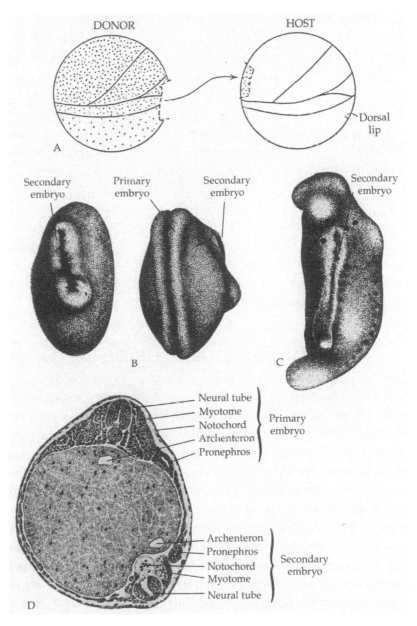

6.1 The dorsal lip transplantation experiment of Spemann and Mangold.
A is a diagram of the operation. B and C show secondary embryos. D
is a cross section showing the structure of the primary and secondary
embryos (B, C, and D are modified from Spemann and Mangold, 1924).
Reproduced from John A. Moore, *Science as a Way of Knowing: The
Foundations of Modern Biology*, Cambridge, MA: Harvard University
Press. Copyright ©1993 by the President and Fellows of Harvard College.

organizers, all of which continued to provide support for organicism. In contrast to Loeb, Spemann did not think it was even necessary to reduce induction to a lower level of organization. Rather, induction was an emergent property and, as such, it could only be fully understood as a tissue-level process. If one looked only at the level of molecules and cells, the process of induction would be lost. Spemann had no doubt that the organizer worked by some sort of chemical or physical process. Certainly molecules were involved, but reducing induction to the level of molecules would not capture the essence of the process. Even working at the level of the cell was inadequate. For Spemann, the most interesting processes occurred at higher levels of organization. He designed experiments that looked at effects of organizer tissue of different ages, of host embryos at different stages of development, and at the ability of organizer tissue from one species to induce secondary embryos in another species.[2] Paul Weiss was impressed with Spemann's work, but also was interested in understanding lower-level processes.

MORPHOGENETIC FIELDS

Paul Weiss began his career in Austria, but later moved to the United States. He agreed with Spemann about the importance of induction and the organizer concept. At the same time he wanted to understand at a deeper level what was responsible for the organization of the embryo. He challenged the idea that the mechanistic and deterministic laws that applied to the inorganic world were completely applicable to biology. At the same time, he made use of both bottom-up and top-down approaches in his experiments. Virchow had coined the phrase *Omnis cellula e cellula,* but Weiss also thought **Omnis organisatio ex organisatione** (All organization from organization) was equally important. An organism was not just the product of cells dividing, but rather was shaped by the organization of the whole developing embryo. Trained in both engineering and biology, Weiss was a pioneer of systems theory and thought the comparison of living organisms to machines was fundamentally misguided. "Machines are passively assembled from parts; living systems actively build themselves, by subdividing a whole cell. In the

---

2 Allen, "Mechanism, Vitalism and Organicism," 277–78.

[living] system, the structure of the whole determines the operation of the parts; in the machine, the operation of the parts determines the outcome." He viewed cell theory through the lens of systems theory and found the reductive approach of Loeb inadequate. His dissertation, in fact, was based on experiments that attacked Loeb's theory of tropisms. Loeb had argued that tropisms (the turning of all or part of an organism in a particular direction in response to an external stimulus) could be understood entirely by reducing them to physics and chemistry. Weiss maintained that many biological explanations were not really explanations, but rather were "nothing but translations of descriptions of facts into inorganic terminology."[3] As others before him had claimed, while the functioning of higher levels of organization such as an organ depended on the proper functioning of the tissues and cells within it, this did not mean that it could be reduced to its lower levels. Each level of organization should be regarded as its own unit and had its own laws.

Weiss did research in a variety of different areas that all provided evidence for his underlying commitment to organicism. These included experiments relating to the nervous system, regeneration, tissue culture, and fiber properties relating to organization and cell biology. Like Morgan, he realized that regeneration and development "are fundamentally of the same nature and follow the same principles."[4] But unlike Morgan, he did not abandon development as being unsolvable at the present state of knowledge and technology. Instead, he thought that applying systems theory would eventually yield results. Regeneration depended on being able to tap into the original powers of growth, organization, and differentiation in the developing embryo. But what were those principles? Did the organism keep a supply of cells in reserve that were totipotent, or could already differentiated cells reorganize themselves when they found themselves in an altered environment? A variety of different experiments on limb regeneration demonstrated that the **blastema** (the mass of cells capable of growth and regeneration into organs or body parts) was not a collection of differentiated cells gathered from different old tissues. Instead, it was

---

3 Paul Weiss, 1925, quoted in Haraway, *Crystals, Fabrics, and Fields*, 148.
4 Paul Weiss, 1939, quoted in Haraway, *Crystals, Fabrics, and Fields*, 161. For original, see Weiss, *Principles of Development*.

a mass of equivalent cells that then differentiated in different directions. Calling the blastema a "herd of organized cells," however, Weiss asked how did this collection of cells obtain the necessary organization to build the particular organ? Embryological writings were often still largely descriptive, inventing terminology to explain what was happening. For example, one might read "the limb field was a property of the field district as a whole and was not associated with a particular group of elements.... The emergence of a definite structure of a tissue or organ meant the composite units ... assume patterned space relations. These reveal themselves in geometric features of position, proportion, orientation ... The ordering processes involved are various referred to as 'organization,' ... 'field action,' and the like."[5] However, as Weiss pointed out, the terms "field" and "organization" were not explanations. What gave the field its particular properties? After World War II, Weiss turned his attention to the mechanism of movement, orientation, and selective contacts of cells.

Trying to work at a deeper level of analysis, Weiss returned to the study of cartilage development. Precartilaginous blastemas of chick limb buds (3–4 days) and of chick eye cartilaginous scleral rudiments were disassociated by means of trypsin. The cells were then allowed to settle out and reassociate in a liquid culture. The new cell clumps were then cultured on plasma clots. Each group developed into true cartilage, but developed with the specific architecture as if they were in a developing embryo. The different cell types had distinctive morphogenetic properties, "which would determine the particular pattern of cell grouping, proliferation, and deposit a ground substance that would eventually develop into cartilage of a distinctive and typical shape." The ability to reorganize into a specific organ or tissue was not a property of single cells, but was rather a group phenomenon. This supra-cellular cell ordering corresponded to the idea of a field. While these experiments did not yet reveal a mechanism, they did show that a field phenomenon could be produced *in vitro*, not just *in vivo*, and would allow further experimental manipulation. For Weiss, the field concept was an organizing principle for all of embryology.

Weiss continued to do experiments in a variety of different systems, which all gave the same result. Cells that had became functional organs,

---

5 Paul Weiss, 1958, quoted in Haraway, *Crystals, Fabrics, and Fields*, 163–64. For original, see Weiss and Moscana, "Type-Specific Morphogenesis," 238–46.

even after complete isolation and disassociation with random recombination, nevertheless reconstituted themselves into the same organ. Doing these experiments *in vitro* demonstrated that cells "know" how to make the correct organ, even in an environment where they aren't receiving any cues. He considered this phenomenon even more fundamental than induction, which had been receiving the lion's share of research attention. He postulated that this cell-specific aggregation might be due to subtle stereochemical surface differences. Interestingly for our story, he drew analogies between this higher level of self-organization and the self-assembly in microtubule synthesis. As we will see in chapter 7, much of Daniel Mazia's work was concerned with the role of **microtubules** in mitosis and cell division. However, it is important to realize that it was also just an analogy. Weiss thought the next step in understanding the study of self-organization into fields would be to examine the properties of single cells that made the higher-level phenomena possible. But again, this did not mean one was reducing the complex to the simpler. Rather, one had to understand how lower-level mechanisms placed *limitations* on the amount of organization possible. There was a higher degree of self-organization that emerged that went beyond just self-assembly. His research on connective tissue in amphibian larvae demonstrated that both top-down and bottom-up approaches were needed to understand tissue and organ formation.

Weiss did not deny that some higher-order complexes could arise from a stepwise interaction of lower-order components. But it was a long way to jump from the self-assembly of collagen fibers *in vitro* to think that even a single cell could be artificially synthesized, much less a whole organism. He viewed the cell through the lens of systems theory. The cell as an independent, isolated, autonomous unit becomes less and less important in such a conception, but instead is subordinated to the system of the organism. Furthermore, even in a single cell, simultaneous synthesis was occurring, not just stepwise processes. Weiss thought that the reductionist understanding of the cell depended on an inappropriate mechanical metaphor that made a split between process and structure. The cell was a dynamic system, and many processes were occurring in a definite pattern, even in the absence of compartmentalization. At the same time, coordination existed among many processes, some of which were segregated and localized. But what was responsible for this coordination?

Microscopic pictures of the reconstruction of the membrane of lamella (small plate of tissue) made of fibroblasts after it had been wounded showed that first epidermal cells migrated and covered the wound. Small fibers appeared between the underside of the epidermis and the fibroblasts. The fibers were randomly oriented, but then a "wave of organization spread over the mass," straightening and orienting them. They then became packed into the characteristic layered structure of the lamella. Again, emphasizing Weiss's organicist perspective, he wrote that this demonstrated the "emergence of a higher order regularity from pre-formed macromolecular complexes, rather than from molecular solution." He continued, "it is the type of principle for which we have as yet no proper explanation in terms of lower-order events."[6] The research on regeneration as well as the remarkable *in vitro* reassociation experiments demonstrated the same principle. The emergent organization was only observed when the components had at some previous point been connected with an *already* organized system. *Omnis organisatio ex organisatione.*

Another person who thought the morphogenetic field was the central concept of embryology was the British biochemist Joseph Needham (1900–95). Needham was one of the most brilliant intellectuals of the twentieth century, with wide-ranging interests that went far beyond biochemistry. Those interests also shaped his approach to biology. A Christian socialist, he thought the logic and philosophy of dialectical materialism was preferable to the static systems of mechanism and vitalism. The organic mechanism as espoused in the writings of the philosopher Alfred North Whitehead also played a significant role in his thinking.[7] He is today most often cited for his pioneering work on the history of Chinese science and is considered one of the greatest Sinologists from the West. In 1954, along with an international team of colleagues, he began a massive project on the science, technology, and civilization of ancient China. The project is still ongoing and to date has produced seven volumes in 27 books. However, before his career took this radical change in direction, he had already established himself in the 1930s as a first-rate embryologist for his research on the biochemical agencies that were responsible for cell differentiation.

----

6 Paul Weiss, 1957, quoted in Haraway, *Crystals, Fabrics, and Fields*, 170. For original, see Weiss, "Macromolecular Fabrics and Patterns," 11.
7 See Haraway, *Crystals, Fabrics, and Fields*, 15.

Needham, like Weiss, advocated an organicism that represented both top-down and bottom-up approaches. For Needham, the solution to the problem of pattern formation would only come about by the unification of biochemistry and morphogenesis. He had been influenced by D'Arcy Thompson (1860–1948) and the necessary role of mathematics to understand the laws of form.[8] Yet he also believed an exclusively mathematical treatment was inadequate. Only by closing the gaps between the level of molecules and that of gross morphological form would development be fully understood. Organization clearly existed at each level. As he claimed, "form was no longer the perquisite of the morphologist, and molecular exactitude no longer the preserve of the chemist." Furthermore, Needham thought the debate over whether life could be reduced to physics and chemistry was fundamentally misguided and a waste of time. Instead, he regarded the task to be elucidated as uncovering the regularities and principles that occurred at each level of organization. They were not reducible to lower levels nor applicable to higher levels.[9]

Needham had also been influenced by C.M. Child's notion of physiological **gradients**. Child had devoted his career to studying development, trying to understand what was responsible for cell differentiation and individuality. Building on the concept of polarity that was first suggested in the work of Boveri, Sachs, and Morgan, he observed that regeneration occurred in a graded process along the axis of the organism, the distinct characteristics seemingly determined by their location. He postulated the existence of specific physiological factors that guided regeneration and was convinced that they could be quantitively measured. He thought the environment was responsible for inducing a metabolic gradient, which in turn provided the foundation for the physiological integration of the whole organism. This was what was responsible for cell differentiation. Many people had criticized Child's theory of gradients for being too general. Ross Harrison, in particular, did not think the concept was useful, in that it did not really specify how various structures were formed. Furthermore, there was not good direct evidence for metabolic gradients. However, the primary goal of Child's research program was not to actually isolate specific factors. Rather, it

---

8 Thompson, *On Growth and Form.*
9 Haraway, *Crystals, Fabrics, and Fields,* 45.

was to provide measurable and quantifiable evidence that supported his theory. He thought these quantitative differences were enough to explain the qualitative differences within the organism.[10] His research emphasized the importance of the environment, and in that way it minimized the importance of both genetics and the role of individual cells, again going against the dominant research paradigm of the time. Needham was more sympathetic to Child's research program, finding the concept of gradients appealing in his quest to understand the exact chemical nature of the organizer. He thought that the organizer was acting something like a hormone and speculated that it was somehow connected to Child's chemical gradients. Needham's later work focused on trying to find hormone-like molecules and the metabolical characteristics of the organizer region. For Needham, structure and physiological activity were inextricably interconnected. Furthermore, differentiation even at the level of molecules was shaped by the whole organism. As Driesch had shown many years earlier, what a given part will develop into depends on its position in the whole organism.

Conrad H. Waddington was another biologist with wide-ranging interests extending to philosophy, poetry, and the arts. He was even Squire of the Cambridge Morris Men, a Morris dancing team that he led on tours throughout the south and southwest of England. Originally trained in geology, he had a multifaceted career, doing research in development, genetics, and theoretical biology. Early in his career, he was able to spend six months in Spemann's lab, which influenced his own embryological research. He was able to demonstrate the presence of organizers in various different tissues of mammals and birds. He found that depending on what kind of manipulations were done and at what stage determined what structures actually formed. Just as Child's work had indicated, positional affects of specific tissue mattered. In the 1930s, he worked closely with Needham trying to further clarify the field concept. He introduced the idea of the individuation field, which was a field associated with the formation of a specific organ with a characteristic shape. How processes were related in both space and time was the key idea of individuation. For example, if a field was cut in half, each half could reform into a complete field that resulted in the formation of two whole organs, but often mirror images of each other.

---

10 Sunderland, "The Gradient Theory."

If two fields were brought together and allowed to fuse, they might rearrange themselves into a single field. If part of a field was removed, often the remainder might be able to compensate and become complete. Sometimes the part that was removed could also become a small, but complete field. In addition to time and three-dimensional space, Waddington thought that concentrations of specific chemicals were important in defining the properties of the field.

However, the concept of fields fell out of favor after World War II for several reasons. First, techniques were still not adequate to analyze properties such as lens induction and limb development. Second, Needham probably did not help his cause by repeatedly quoting (sometimes inaccurately) communist theoreticians such as Engels and Lenin, as well as Soviet ideologues who supported a nonreductionist organicism. Third, a lot of this research had been done in Europe, particularly in Germany, and the scientific infrastructure had been badly damaged as a result of the war. Finally, and perhaps most important for our story, is the dominance of genetics in both evolutionary theory as well as development. Morgan and other geneticists saw the concept of fields in direct opposition to their explanation of heredity. They actively campaigned to keep research on fields by investigators such as Child and his students from being published.[11] With the success of genetics, and the emergence of molecular biology as the dominant paradigm for doing research in biology, the concept of morphogenetic fields still seemed too vague – overly holistic to the point that there was no way to subject it to biochemical or genetic analysis. Like organicism, fields were just another example of an ill-defined scientific concept.

GENES AND DEVELOPMENT

For the most part, geneticists were still not that interested in development. An important exception was the geneticist Curt Stern. He claimed the early patterning of the embryo was due to embryonic fields and that genes were responsible for when and where the fields were expressed. "The pre-pattern of the embryonic tissue in *Drosophila*,

11 Gilbert and Sarkar, "Embracing Complexity," 5.

which calls forth a response of genes involving the differentiation of bristles, are embryonic fields of larger dimensions than the limited points of normal location of bristles."[12] Stern even hypothesized that fields were products of genes. However, this also meant that fields were merely epiphenomena of genes. Even if fields were real, they were only byproducts of genes, according to Stern.

Just as most geneticists were not interested in how genes affected development, neither were most embryologists. The exceptions were people who thought broadly and crossed disciplines, and Waddington was one such person. He certainly did not think that fields were merely epiphenomena of genes, but he also recognized that understanding gene action was critical to understanding development. His concepts of **canalization**, the **epigenetic landscape**, and **genetic assimilation** have continued to influence research in development and demonstrated the necessity to reunite heredity with development to understand morphogenesis. Although he did much important work on morphogenetic fields, perhaps his lasting legacy will be in this later research. Not only did he want to integrate genetics with development, he also wanted to integrate development with evolution.

Canalization refers to the property of developmental systems to produce the same phenotype, in spite of environmental or genetic perturbations that might otherwise disrupt development. It is a buffering system that protects the developing embryo from disturbances. Furthermore, canalization allows for genetic variability to build up, although it is hidden, since it is not expressed in the phenotype. Closely tied to canalization was Waddington's idea of the epigenetic landscape, which illustrated the different developmental pathways a cell could take in the process of differentiation and organogenesis. He envisioned the landscape as a series of valleys and ridges. The cell was a ball starting at the top of the landscape that rolled down to the bottom, which represented the final differentiated state of the cell. Each valley represented a particular developmental pathway the cell could take (see figure 6.2). It would tend to move to the lowest point in the center of the valley, but it could move up the sides of the valley as a result of either environmental or genetic perturbations. However, because of canalization, represented by the

---

12 Curt Stern, quoted in Beloussov, "Life of Alexander G. Gurwitsch," 778.

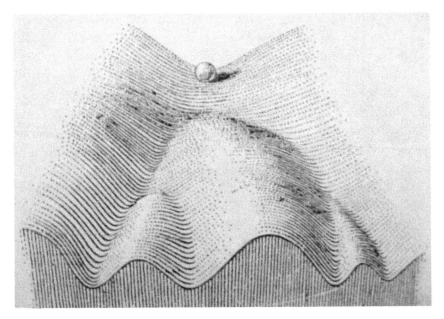

6.2 The epigenetic landscape. Image originally published in 1957. Reproduced from C.H. Waddington, *The Strategy of the Genes RLE*, 1st edition (London: Routledge, 2014), © reproduced by arrangement with Taylor & Francis Books UK.

height and steepness of the ridges of the valley, the ball will tend to be pushed back to the center and continue down the same pathway. The steepness of the slope of the sides of the valley represented the tolerance or the amount of buffering that insured the same phenotype, and also symbolized the ability of the genes to regulate differentiation. There are various branch points at the beginning of the hills that represent points of disequilibrium, where induction and tissue interactions can push the ball (cells) down into a different valley. Waddington thought that genes generated the inducers and that the formation of an eye or a limb could be analyzed by mutation in the genes that encoded the inducing molecules. These processes thus are regulated by genes.[13] The epigenetic landscape was a model that illustrated the interaction between genes and the environment that determined the developmental pathways that undifferentiated cells could

---

13 See Hall, "Waddington's Legacy," 113–22.

take, eventually leading to fully differentiated tissues. Canalization and the epigenetic landscape suggested a nonvitalist way to explain Driesch's concept of entelechy.

Driesch borrowed the term "entelechy" from Aristotle, using it to describe the developmental tendency to completeness or perfection in both normal and experimentally manipulated conditions. Because of its vitalistic overtones, not only it but also the organicism that was associated with it were discredited. But the fact of the matter is entelechy is a real phenomenon. Driesch used it as an *explanation* for biological phenomena rather than as something that itself needed to be investigated.[14] It is still not completely understood. Waddington's epigenetic landscape was a model that suggested a means of investigation to explain how entelechy actually occurred. Canalization is an aspect of the more general phenomenon of biological homeostasis. It is a phenomenon that can't be explained either at the level of molecules or at the level of the cell. It is an "emergent" phenomenon (such as temperature and semipermeable membranes) that can't be explained by the known properties of its component parts. In the organicism that he espoused, Waddington also realized that genes played essential roles in morphogenesis.

Waddington claimed that his model could explain evolutionary adaptation as well. He continually criticized population genetic models as inadequate to explain how genes were really acting at the level of development. And he insisted that the evolution of organisms was really the evolution of developmental systems. He maintained that the embryo had the genetic ability to respond to different environmental disruptions. Genetic assimilation is the process by which a particular phenotypic character, which was initially produced in response to an environmental influence, then becomes canalized through a process of selection. The trait is still formed, even in the absence of the environmental influence. How big a role genetic assimilation plays in evolutionary adaptation remains controversial. Although more and more examples continue to accumulate in its favor, it is extremely difficult to evaluate if a new adaptation is due to this process or any of a variety of different mechanisms, particularly the primary one that most evolutionary biologists still espouse: mutation and selection.

---

14 Garcia-Bellido, "Cellular Interphase," 8.

TENSIONS

Each of the biologists in this chapter in their own way were investigating the kinds of questions that most interested Huxley. Indeed, Needham even quoted Huxley in the introduction to his book *Chemical Embryology*: "The goal of physiology was to deduce the facts of morphology ... from the laws of the molecular forces of matter."[15] The organicism these scientists espoused was free of any vitalist implications and emphasized the need to understand development using both top-down and bottom-up approaches. By regarding morphogenetic fields as the central and unifying concept of embryology, they emphasized levels of organization that were higher than the cell and thus that the cell couldn't be considered independent. At the same time they recognized that one had to understand what was happening at the level of molecules. They may not have explicitly criticized cell theory, but their research certainly demonstrated the limitations of it in trying to understand development. This research also highlights the ongoing tension that existed between two different approaches to understanding morphogenesis, even within an organicist framework that recognized the need to work at many different levels of organization.

Lewis Wolpert described this tension by characterizing the community of developmental biologists as having a left wing and a right wing. The right wing, following in the tradition of Weismann and Wilson, thought that pattern formation and the organization of spatial differentiation was determined and internal to the cell. For them, cell interactions played a very small role. Autonomous cell lineages and cytoplasmic localization were responsible for cell differentiation and generating diversity. The left wing, following in the tradition of Driesch, thought gradients and more global interactions were what were primarily responsible for differentiation. Then there were the people who were in the middle, such as Hertwig and Spemann, who thought local interactions such as induction were the most important. All of these distinctions were important but, as Wolpert wrote, he hoped that the right, left, and middle can all come together. None of them needed to be mutually exclusive. There is supportive evidence in favor of all of

---

15 Thomas Huxley, quoted in Needham, *Chemical Embryology*, 9. For original, see Huxley, "The Study of Zoology," 83.

them. For our story, all of the evidence supports the necessity of taking a more holistic approach to understanding development. In doing so, it suggests a rethinking of the role of the cell as espoused in classical cell theory.

These ideas in no way minimize the importance of cells. Yet if we return to the most fundamental problem in all of biology that remains unexplained, it is the origin of life itself. Darwin titled his book *On the Origin of Species by Means of Natural Selection,* but really a more appropriate title (although not nearly as catchy) would have been "How Species Arise From Other Species." To be fair, he did explain this in the subtitle: *Or the Preservation of Favored Races in the Struggle for Life.* Just as Darwin began with already existing species to show how they could be transformed, embryologists began with already existing cells in trying to understand development, regardless of what level of organization they were working at. Cell theory made a major contribution to embryological research by showing that all life was made up of the same basic structural unit and, furthermore, that life did not spontaneously arise. Yet at some point in the distant past, life did emerge from a particular arrangement of lifeless molecules that formed a structure that was "alive."

Progress in understanding how life first arose has been made by recognizing that one cannot reduce life to just the idea of a replicating molecule. Rather, metabolism was also a crucial component of what we call life. In addition, since a key aspect of something being alive is that it reproduces, a very important question is what determines when a cell decides to divide? Fully understanding the answer to this question has profound implications for development. The next chapter explores the work of cell biologist Daniel Mazia (1912–96), who we have already briefly met, and who devoted his career to trying to answer this most fundamental question.

CHAPTER SEVEN

Twoness

The cell cycle can be viewed as a bicycle with a Reproductive Wheel and a Growth Wheel.

Daniel Mazia, "Origin of Twoness," 1978

MORE THAN DNA: DANIEL MAZIA AND THE IMPORTANCE OF CELL BIOLOGY

DNA went from being regarded as a "stupid molecule" to the master molecule that held the secret to life. While progress in our understanding of genes and gene action continued to garner the most attention in the popular press as well as in historical treatments of biology in the last part of the twentieth century, important breakthroughs were occurring in other disciplines as well. Advances in technology, especially the development of electron microscopy, meant that the cell was revealing more and more of its secrets. Cell biology was becoming a recognized discipline in its own right. Indeed, to the cell biologist, DNA was as much the controlled as the controller. Enzymes were telling it when to replicate, and the mitotic spindle was separating those passively pulled strands of nucleic acid. There have always been scientists who have realized the limitations of a strictly reductionist research program, but in modern biology their achievements have often been in the shadow of the dominating paradigm of genetics and molecular biology. Perhaps no one more clearly showed the limitations of the strictly reductive approach than the person who first isolated the **mitotic apparatus** – Daniel Mazia. In Mazia's work, we continue to follow the thread of the kinds of issues that Huxley initially raised with his critique of cell theory, as well as see how advances in microscopy continued to play a critical role in our understanding of the cell. Mazia can rightly be considered one of the founders of modern cell biology, and his provocative ideas about the organization of the cell continue to influence researchers today.

Daniel Mazia was born on 18 December 1912 in Scranton, Pennsylvania, and grew up in Philadelphia. The son of Russian Jewish immigrants who ran a corner grocery store, the prospects of becoming a cell biologist during the Depression did not seem a very promising career choice. Who even knew what a cell biologist was in those days? But when a close friend died of cancer while he was attending the University of Pennsylvania, he became interested in unraveling the fundamental processes that ran amuck and made cancer cells start wildly dividing. He went on to receive his PhD at the university in 1937, working with the calcium ion pioneer L.V. Heilbrunn on the role calcium played in fertilization. They also used what had become the organism of choice – sea urchin eggs – to study the early stages of development. After graduate school he was a research fellow at Princeton University and Woods Hole Marine Biological Laboratory. His association with Woods Hole lasted many years as both a teacher and a trustee. In the early 1950s he taught a physiology course that became quite famous. He was at the University of Missouri from 1938 until 1951, and then a professor in the zoology department at University of California, Berkeley until state laws required his mandatory retirement in 1979. However, Mazia was not about to retire and jumped at the chance when the director of Hopkins Marine Station of Stanford University recruited him. He continued to teach and do research there in Monterey, California, until his death in 1996. While there, he also became president of the International Cell Research Organization (ICRO), a branch of UNESCO. In that capacity he continued to inspire young cell biologists in ICRO workshops held all over the world, from Asia to Europe to South America.

Mazia drew from a wide variety of disciplines including biochemistry, biophysics, embryology, genetics, botany, and zoology in his investigations of the cell. As one of his former students wrote, "It's impossible to separate Professor Mazia from the entire fabric of cell biology."[1] Perhaps his most lasting contribution was the many, many students he trained. His research inspired students who went on to form their own labs with their own students. One of the perks of being Mazia's student at Berkeley was regular field trips to northern California sites to collect sea urchins. More importantly, this provided opportunities for his students to engage with the man in a more informal, but important, way. Before his large lecture course began, he could often be found strolling in the redwood grove of the campus, collecting his thoughts about what he would say in a course he had taught probably hundreds of times, but wanting to bring something new as a way to engage his students. Yet he was always happy to talk to a student who approached him.[2] As his former student at Berkeley and then colleague at Hopkins Marine Station, David Epel wrote, "He never looked at anything conventionally. He always had different ways of looking at the same problem.... He spoke like literature. It was not just a transmission of facts; it was stimulating. He spoke to ideas and images in his listener's mind, so a student and a senior investigator could listen to the same lecture and gain something different. It was like a reverie you would come away with your mind abuzz with new ideas."[3] And he was more than willing to poke fun at himself when some of his ideas went astray: "I thought what I thought because I thought without thinking."[4] He was concerned with understanding life at the most fundamental level and brought a deeply philosophical approach to his life in the laboratory. As his colleague wrote, "He had a way of looking at cell biology that went beyond the current preconceptions to fundamental issues. He could see things that no one else could see clearly, and he inspired others to do the same."[5]

---

1 Gerald Schatten, quoted in *Stanford News*.
2 Personal reminiscence of the author.
3 David Epel, quoted in *Stanford News*.
4 Daniel Mazia, quoted in Epel and Schatten, "Daniel Mazia," 417.
5 Richard Steinhardt, quoted in *Stanford News*.

MITOSIS AND THE CELL CYCLE

Mazia is best known for his work with the Japanese biologist Katsuma Dan in isolating the structure responsible for moving the chromosomes as an integral body, which he named the mitotic apparatus (MA). This isolation ended once and for all the caveat that the spindle observed in cells during mitosis might have been a fixation artifact. Studies by him and his students were focused on explicating the structure of the MA and the general problem of cell reproduction. The MA consists of the centrosome and spindle-like fibers made up of microtubules. During cell division these fibers attach to sister chromatids of a chromosome, which then separate and ensure that each daughter cell has the exact correct copy of each chromosome (see figure 7.1). He was a wonderful writer; when Jean Brachet and Alfred E. Mirsky asked him to contribute a chapter to their classic edited six-volume *The Cell*, it ended up as separate volume with only one other chapter on meiosis. Mazia wrote what they referred to as "a magnum opus," and they had "no desire to cut such a fine treatise on mitosis and meiosis."[6] Mazia's group isolated an important protein in the mitotic apparatus that eventually was identified as tubulin and turned out to be absolutely critical to the various structures made up of microtubules in the cytoplasm. Continuing research provided great insight into cellular organization, including that of the cytoskeleton, membrane organization, and the dynamics of chromosome movement. Chromosomes contained the hereditary material, the genes, but Mazia was not a geneticist. As became clear with the isolation of the mitotic apparatus, he quipped in a wonderful descriptive statement of mitosis that "the role in mitosis of the chromosome arms, which carry most of the genetic material, may be compared with that of a corpse at a funeral: they provide the reason for the proceedings but do not take an active part in them."[7] Mazia's interest lay in understanding exactly what those proceedings were. This in turn gave rise to a host of other topics that essentially defined the agenda for cell biology for decades to come. This included macromolecular control in cell proliferation, regulation of the cell cycle, control of genetic expression, microtubule assembly, and the structural and molecular basis of cell movement and cell differentiation. But as Mazia recognized, these various processes

---

6   Brachet and Mirsky, *The Cell*, vol. 3, vii.
7   Mazia, "Mitosis and the Physiology," 212.

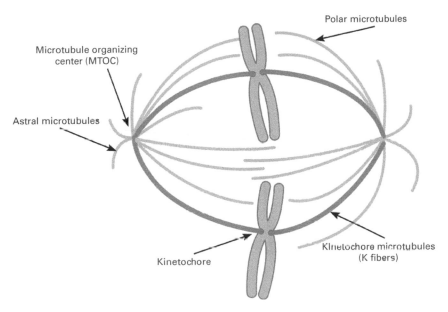

7.1 The mitotic apparatus. By Lordjuppiter, CC BY-SA 3.0, https://
commons.wikimedia.org/w/index.php?curid=17690038.

could be considered at their most fundamental level contributing to the
"origin of twoness."[8] Early researchers thought of the cell as having two
phases – the resting phase R and the mitotic phase M. It is not surprising that so much early work was directed at mitosis, since that could
be seen and was quite dramatic. No chromosomes were visible in the
resting stage or interphase, and it didn't seem like much was going
on. In spite of working on mitosis, Mazia emphasized that events of
the mitotic cycle were not only what was observed during mitosis, but
ran through the whole cell cycle. By the early 1950s it had been shown
that the duplication of the chromosome material occurred during
interphase. Interphase was divided up into three phases: G_1, S, and G_2.
Normal growth and metabolism, including synthesizing compounds
that would be needed for duplication of DNA, occurred during G_1. In
the S phase the DNA was duplicated. In G_2, further growth occurred
and other molecules were made to prepare the cell for mitosis and cell

---

8 Mazia, "Origin of Twoness," 3.

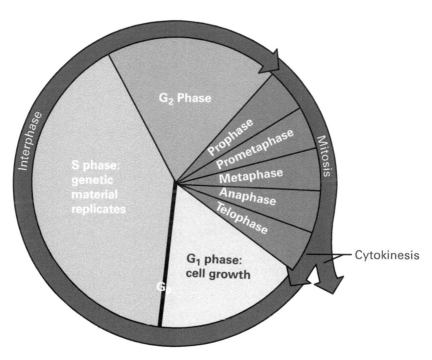

Interphase

G$_2$ Phase

Prophase

Prometaphase

Metaphase

Anaphase

Telophase

Mitosis

S phase: genetic material replicates

G$_1$ phase: cell growth

G

Cytokinesis

7.2 The cell cycle

division (see figure 7.2). This basic scheme is still taught today and is certainly useful. Mazia did not reject it, but it was an idealized scheme that he thought was more useful for students "in the history of the cell, as other historians invent a Renaissance fully inhabited by Renaissance Men (but not women) or Enlightenment in which people were enlightened."[9] However, who can say whether cells with this cycle outnumber ones that don't have it? Some cells, such as blastomeres in embryos, do not have to grow in order to divide; others such as lymphocytes have to more than double their size before they divide. Some cells have no G$_1$, others no G$_2$. Mazia impishly pointed out to his academic colleagues that "the typical form of the cell cycle does apply to the newly evolved phylum of animals Phylum Falconia: animals which use plastic dishes as shells and feed on research grant funds."[10] Mazia was poking fun at his fellow scientists who were growing all their cells in plastic dishes from

9 Mazia, "Origin of Twoness," 3.
10 Mazia, "Origin of Twoness," 3.

Falcon, the leading manufacturer of plastic tissue culture lab ware. At the same time, Mazia was implying that this was quite an unnatural environment to be growing cells. Thus it shouldn't be surprising that these Falconia cells might not follow the typical stages of the cell cycle. Rather than emphasizing the timing of events and how common this cycle was, Mazia was interested in considering the necessities for one cell to become two. He was trying to isolate the events in the cell cycle that were specifically involved in reproduction, distinct from the myriad activities that a cell performs when it is not dividing. In other words, he was interested in the origin of twoness. He suggested that the cell cycle could be thought of as a bicycle with a reproductive wheel and a growth wheel, with the two wheels geared together in different ways in different cells. He thought that the sequence of events that constituted the reproductive wheel could be defined quite sharply, and that none of the events or molecules of the reproductive cycle were needed by the cells unless they were actually progressing toward division.

Different cells exhibit a lot of variability in their cell cycle, but from all the different cell types it became clear that the turn of the reproductive wheel was inherently rapid. Thus to understand how regulation of the cell cycle occurs, the answer would be found in the processes of the growth wheel and how they *retard* the reproduction wheel. For example, the *Drosophila* egg's complete reproductive sequence from mitosis to mitosis is about 15 minutes. In such cases the egg must be fully provisioned for these first divisions, and very little is needed to be synthesized to actually initiate duplication of the chromosomes. Sea urchin eggs took 30 to 35 minutes after fertilization to divide, suggesting that certain compounds must be synthesized. Mazia had isolated some proteins that needed to be synthesized, and Tim Hunt, not knowing of this work, independently discovered these proteins, which he named "cyclins." Mazia maintained that the broader take-home message from this work was that the regulation of the cell cycle was the result of limitations or retardations, the slowing down or blocking of an intrinsic 10-minute cell cycle. Therefore, some seeming inconsistencies in findings on cell regulation really shouldn't be considered problematic because virtually anything that growing cells do could act as a limiting factor. Furthermore, one would expect that different types of cells would have different regulatory signals.

In earlier work Mazia had referred to growth as "friction" in the cell cycle. In a colorful analogy, he claimed it was useful to know that feathers and bricks fall at the same rate in a vacuum, but we had better go to Pisa and learn how one falls more slowly than the other in the real world. Likewise, in the real world of actual cells, the kinds of events that could limit the reproductive cycle were almost infinite. Beans take 19 hours for the complete cycle. Most nerve cells lose their capability to divide once they reach maturity, while liver cells retain, but do not normally utilize, their capacity for division. This idea of a slowing down or retardation had all kinds of implications for development and also when regulation went awry, as in cancer. Mazia's research was premised on this core idea: the chromosome cycle runs throughout the entire cell cycle. What was seen in mitosis represented only the final orders of folding and coiling of the chromosome.

In later work Mazia suggested that it was the decondensation of the chromosomes in the first part of interphase that determined whether the cell will go toward replication; other findings suggested that the mechanisms of decondensation were the levers in the control of the cell cycle.[11] Nevertheless, he thought that deeper problems remained – in particular, understanding the packing at a three-dimensional level with every genetic segment in place as well as the positioning of the **kinetochores** (a protein complex that forms during mitosis and where the spindle fiber attaches to the **centromere** region of the chromatids). This would be critical to explicating how one cell became two. Furthermore, he thought one of the deepest components of the living cell was time. A variety of research suggested that there were underlying cycles in the cell that drove events expressed in both the chromosome cycle and the cell cycle. Could those cycles also possibly be influenced by the organism as a whole?

In addition to chromosome duplication, the second aspect of how one cell becomes two was the actual splitting and making of two cells. This was a result of bipolarization with the establishment of the mitotic poles or centers. The poles appeared to be made by the centrosome, which consisted of smaller bodies that were called **centrioles**. The poles must reproduce, and these sister centers must split apart. This was accomplished by the growth of spindle fibers (which were made of microtubules) that attached to the sister chromatids and pulled them apart. This scheme required that the mitotic centers not only must be

---

11 Mazia, "The Chromosome Cycle and the Centrosome Cycle," 62.

capable of reproducing, but also serve as a special kind of **microtubule organizing center (MTOC)**. Centrioles provided a basis for thinking about properties of the mitotic poles, but they created their own set of problems. Mazia pointed out that "cell biologists are morphologists at heart and unlike in physics we want to see a particle when we think of a pole and if it is supposed to reproduce we expect to be able to see it budding off."[12] Unfortunately, many times the centrioles couldn't be observed, and in some organisms they appeared and disappeared. Furthermore, the functions of the poles seemed to be the same whether centrioles were seen or not. New evidence showed that the activities of the poles did not depend on organized centrioles, even when they were identified. Centrioles and mitotic centers of maternal origin could not be found in eggs, but were contributed by sperm upon the fertilization of the egg. (As an aside, Mazia pointed out this justified the existence of fathers.) However, some tricks of artificial **parthenogenesis** (the development of the organism in the absence of fertilization by sperm) can bring about the presence of "cytasters" in unfertilized eggs, and they can function as mitotic poles. Sometimes centrioles were found in those induced cytasters. Because of this, Mazia maintained mitotic centers were real; they were not abstractions to explain mitosis in the absence of centrioles. The problem then was to understand how these centers made centrioles, rather than the other way around. Indeed, Mazia claimed that "[n]othing we have learned about mitosis since it was discovered a century ago is as dazzling as the discovery itself."[13]

THE CENTROSOME AND THE CELL CYCLE

In 1883 Edouard van Beneden (1846–1910) identified definite corpuscles while observing mitosis in the eggs of a roundworm parasite of the horse. In 1888 Theodor Boveri described them in greater detail and named them centrosomes. Boveri realized their significance, writing a massive 200-page treatise entitled "On the Nature of Centrosomes." In it he claimed that the centrosome was an "autonomous permanent organ of the cell ... the dynamic center of the cell ... the true division-organ

---

12 Mazia, "The Chromosome Cycle and the Centrosome Cycle," 72.
13 Mazia, "The Chromosome Cycle and the Centrosome Cycle," 49.

of the cell.... It coordinates nuclear and cytoplasmic division."[14] The fusion of the egg and sperm initiated development in most animals, each contributing half of the hereditary material. However, Boveri recognized that the egg and sperm also made different contributions:

> The ripe egg possesses all of the elements necessary for development save an active division-center. The sperm, on the other hand, possesses such a center but lacks the proto-plasmic substratum in which to operate. In this respect the egg and sperm are complementary structures; their union in syngamy thus restores to each the missing element necessary to further development. Accepting this, it follows that the nuclei of the embryo are derived equally from the two parents; the central bodies [centrosomes] are purely of paternal origin; and to this it might be added that the general cytoplasm of the embryo seems to be almost wholly of maternal origin.[15]

However, Jacques Loeb disagreed with Boveri on the role sperm played in fertilization. In his experiments with artificial parthenogenesis, he had shown that changes in ion concentration could excite the unfertilized egg and stimulate development. He thus minimized the importance of the role sperm played. Nevertheless, not only did sperm contribute chromosomal material, Boveri had shown that the centrosome provided the apparatus for the mitotic poles. He suggested that the egg also had a centrosome that became disordered, but not destroyed, as the egg matured. He hypothesized that it was resurrected by parthenogenetic procedures in the absence of sperm, creating a mitotic pole that then allowed development to occur.

Unfortunately, at the turn of the twentieth century interest in the centrosome went out of vogue when cytologists could not find evidence of it in many eukaryotic organisms. It was much more satisfying to be characterizing the chromosomes and making chromosome maps that correlated specific areas on the chromosomes to the traits Morgan was producing in his flies. Even Mazia in his 1961 treatise on mitosis discussed the reproduction and function of the centrosome, but said

---

14 Theodor Boveri, 1901, quoted in Mazia, "Centrosomes and Mitotic Poles," 1.
15 Theodor Boveri, 1901, quoted in Schatten and Stearns, "Sperm Centrosomes," 1178.

very little about its actual physical attributes. Much later in a letter to his old friend and colleague Katsuma Dan, he explained why he did not turn his attention to the centrosome earlier, in spite of his interest in cell division. He pointed out that there wasn't much interest in the centrosome when they were students. Cytologists couldn't find it in many cells. Yet as Mazia wisely recognized, seeing may be believing, but what you see is determined in part by what you think you should see. "The defect wasn't in their eyes, but in their erroneous preconception. They were looking for a compact particle."[16]

Mazia continued his investigations, further characterizing the mitotic apparatus and elucidating the details of mitosis, and in doing so came to agree with Boveri about the centrosome's importance. At the same time, in describing the disagreement between Loeb and Boveri over artificial parthenogenesis, he claimed that the two men had each defined only one half of the problem, but thought it was the whole problem, and therefore dismissed the other person's ideas as irrelevant. Mazia took both men's ideas into account and argued that the formation of a mitotic apparatus was essentially a two-step process. Loeb's ideas explained step one, the activation of the egg, while Boveri's hypothesis defined step two, the restoration of the maternal centrosome to make a functioning mitotic pole. Viewing the problem this way, Mazia also elucidated what a true functioning mitotic pole could do. Why did he think that the centrosome was so important? For the same reasons as Boveri – because it was the key to one cell becoming two. The centrosome was the critical component to a fully functioning mitotic apparatus that allowed for the exact partitioning of the hereditary material into two cells.

Mazia claimed that Boveri and later workers until relatively recently had pictured the MA as a kind of puppet theater in which the strings or spindle fibers and the microtubules tied kinetochores to the centrosomes. The connection set up a virtual pull. A lot of work supported this basic idea, but Mazia pointed out that the puppet play existed in a sort of cytoplasmic void where the centrosomes and chromosomes were hanging out, and either the centrosomes or possibly the kinetochores cranked out some fibers from a pool of buffered tubulin that was lying around. Wonderful experiments had demonstrated that isolated centrosomes could capture isolated chromosomes, yet a fundamental

16 Mazia, "Mitotic Poles in Artificial Parthenogenesis," 525.

problem remained: how did the accurate engagement of chromosomes occur? Mazia wanted more scenery and stagehands for the puppet show. He maintained that the roles of the actors needed to be examined more closely, particularly the role of the centrosome: "There could not be many questions more important than: is the centrosome a universal, permanent, reproducing organ of the eukaryotic cell; or is it an agent that can be generated *de novo*?"[17] If it was a permanent universal reproducing organ, this would go a long way in explaining how one cell becomes two, accurately, generation after generation.

Mazia began his 1984 paper on centrosomes and mitotic poles with a quote from Shosaburo Watase to emphasize the significance of the centrosome. Watase was both a student and a lecturer at Woods Hole, had trained with Whitman, and eventually also became a lecturer at the University of Chicago. Whitman said of him, "Watase is the broadest and soundest student of cellular biology in America," even surpassing E.B. Wilson. In 1894 Watase claimed that "Professor Flemming's remark that the discovery of the centrosome marks as important in the history of biological science as did the discovery of the nucleus seems certainly justified."[18] Recall from chapter 3 that in the late 1800s Flemming had discovered that the nuclear material took up stains, which he called chromatin. He had observed mitosis and argued that the nucleus was the carrier of the hereditary material. He had, however, initially disagreed with the importance being attributed to the centrosome. For him to have not only changed his mind but to have claimed that the discovery of the centrosome was as important as that of the nucleus was indeed a very bold statement.

It was not an accident that Mazia began his paper with Flemming's comment, as he wanted to emphasize how "central" the centrosome was to understanding the life of the cell. Mazia made a provocative suggestion. He argued that the idea that the centrosome had a consistent morphology and that it was a particle needed to be abandoned. Rather, it was linear and flexible, not only capable of taking different shapes in different cells, but also many different shapes within the cell cycle.[19] Furthermore, he considered the reasons why the form of the centrosome

---

17 Mazia, "Mitotic Poles in Artificial Parthenogenesis," 525.
18 Shosaburo Watase, quoted in Mazia, "Centrosomes and Mitotic Poles," 1. For original, see Watase, *Biological Lectures*, 272.
19 Mazia, "Centrosomes and Mitotic Poles," 1–15.

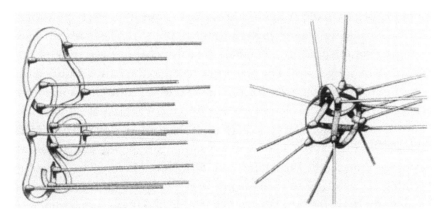

7.3 Model of the flexible linear chromosome. The conformation of the linear centrosome determines the three-dimensional structures built of microtubules. The linear element carries microtubule-generating units, each of which specifies the origin and direction of one microtubule. Left, the centrosome functions as a flat pole, generating a "barrel" form of the half spindle. Right, a compact centrosome generates a half spindle with a pointed pole and an aster. Reproduced from Mazia, 1984.

should vary. He examined the evidence in light of three assumptions: (1) the centrosome did take on different forms, (2) the form varied through the life cycle of the cell, and (3) the centrosome functioned as an MTOC in which the alternative forms determined the shape of the mitotic poles as well as the other organizers, which in turn determined microtubular structures. Compact centrosomes should generate mitotic spindles with pointed ends, and flat centrosomes should generate spindles with flat ends (see figure 7.3). Instead of thinking of it as an organelle, Mazia proposed that a more appropriate comparison would be to a chromosome. It was large, it was present in relatively small numbers, it reproduced only once in a cell cycle, it condensed in one part of the cell cycle, and it then dispersed and even seemed to disappear. The parallels with chromosome behavior were striking. To bolster his argument, he reminded his colleagues that the continuity of chromosomes was disputed for some time, even after the genetic evidence for their permanence was irrefutable. Just as Mazia had indicated, cell biologists were morphologists and many were waiting for techniques that showed centrosomes had not disappeared. Mazia thought that evidence for the permanence of the centrosome would also have to wait for new methodologies in electron microscopy to demonstrate irrefutably what he was suggesting. Interestingly, Boveri had observed in sea urchin eggs compact centrosomes at metaphase that then became flat wide plates as anaphase progressed. The plates divided, and the halves rounded up and formed compact poles for the next cycle. Unfortunately, this idea of a centrosome cycle did not make it into the teaching of cell biology, despite even seeing illustrations of what Boveri had described in textbooks! However, at the time there was significant skepticism of classical cytological methods. Remember how dismissive Miescher had been of cytologists, referring to them as just dyers. In this case seeing was *not* necessarily believing. Mazia's group had repeated Boveri's experiments, further confirming his observations by means of electron microscopy.[20] Mazia recognized the critical function of a true mitotic pole and claimed that some of the literature on artificial parthenogenesis greatly exaggerated its success rate. For example, the procedure in sea urchins rarely resulted in normal development. Instead, a great variety of different pathologies were observed that correlated to different abnormal MAs, which in turn could be traced to abnormal centrosomes. As Boveri had hypothesized, the mature egg did contain a denatured centrosome that could generate a nonpolar MA after step one in Mazia's two-step scheme. Step two renatured the centrosome, but extensive cytological studies showed that the renaturation was most often imperfect, generating a wide variety of MAs. These included not making a pole at all, or making ones that caused the chromosomes to line up in a variety of different ways. "At the least, this mitotic freak-museum would refute any notion of the centrosome as an entity that is either present or not present and is always the same when present."[21] No wonder normal development rarely occurred. Boveri was correct; the centrosome was absolutely essential to normal cell division.

The centrosome had a definite cycle of disappearing and appearing, and also went through characteristic changes of shape and position in the cell. By giving up the expectation that centrosomes must be compact bodies, one could then look to the different forms of the poles to glean information about the various forms of centrosomes. Mazia suggested that he might be charged with circular reasoning but wryly wrote,

20 H. Schatten et al., "Behavior of Centrosomes," 105–09.
21 Mazia, "Centrosomes and Mitotic Poles," 6.

"circles are not necessarily vicious." Although the exact timing and mechanism of how the centrosome replicated itself was still unknown, he made the controversial suggestion that the changes in shape it exhibited during the cell cycle were the actual stages in its division, and thus the making of one cell into two. This was in spite of the fact that the centrosome cycle was not tightly synchronous with the chromosome cycle. It was clear that the centrosome served as a MTOC and initiated the formation of the **spindle apparatus**. Both the shape and its orientation always determined the plane in which the cell divided. Mazia proposed a model of the centrosome in which he hypothesized a linear element, a chain of beads that "determines the arrangement of the microtubules initiating units and the orientation of the individual units determines the morphology of the spindle apparatus," which were built of microtubules. Furthermore, what was critical was the "sideness" of the chromosome that put the kinetochore on one side of the chromosome so that sister kinetochores faced in opposite directions. Their engagement with the poles they faced ensured the accurate separation of sister chromosomes. The generation of the half spindle apparatus also ensured that by the end of the process only one spindle fiber could attach to each kinetochore. This suggested that the centrosome was more than just a MTOC.[22] Mazia made an audacious speculation: perhaps the centrosome contained not only the information for the structure and orientation of the mitotic apparatus, but it contained information about *all* of cell morphology. Work by others had analyzed images of the centrosome obtained by high-voltage electron microscopy and found that the microtubules guided movement of pigments within the cell. They concluded that the centrosome "is an organelle where information about the three-dimensional structure of the cell is stored."[23] Electron microscopy had also revealed that the mitotic apparatus consisted of a mass of membranes that was organized by the poles. How it did this was not known. Mazia was convinced that answering this question needed a different approach than the dominant research paradigm of the time.

While Mazia was impressed with what was beginning to be understood about the exact molecules that made up the centrosome, he thought the

22 Mazia, "The Chromosome Cycle and the Centrosome Cycle," 49–92.
23 Nahum D. Gershon, Keith R. Porter, and Mark A. McNiven, quoted in Mazia, "The Chromosome Cycle and the Centrosome Cycle," 79. For original, see Gershon, Porter, and McNiven, "Three Dimensional Structure," 65–66.

emphasis on identification of molecules avoided deeper questions of centrosome structure and its precise mode of reproduction: "Something truly fundamental is missing in our image of the cell and that 'something' exists at a level of complexity higher than that of molecules and perhaps more comparable to the complexity of chromosomes."[24] He did not discount all the tremendous advances in molecular genetics. Chromosomal genes were transcribed, processed, and translated, generating a wealth of macromolecules, but he thought it was time to find out how molecules make cells. Cell biology had a critical role to play in the interpretation of the chain of command dictated by the genes. He suggested the centrosome might be the organ of interpretation through which the entire structure was managed. For Mazia, centrosomes were potentially far more than just the organizer and initiator of microtubule synthesis. They were "*bearers of information about cell morphology.*" How and when does a cell divide? The answer lies in the interaction between the particular cell and the information-coordinating action of the whole organism. The control of mitosis may be viewed from the inside looking out, or from the outside looking in. Mazia's research was guided by the idea that "the clear questions and answers will be found where these two views meet."[25]

THE CELL BODY

Until his death, Mazia was working on an idea that he called the cell body, a kind of supramolecular organization of the entire intracellular structure. It was generated by a nuclear/centrosome complex that included the microtubule assemblage. The cell body pervaded the whole interphase cell and condensed into a mitotic apparatus during mitosis. The shape of the cell body would be determined by the arrangement of the microtubules and other structures within the cytoplasm that were associated with the microtubules. Evidence for the existence of cell bodies had been found in developing embryos of both *Drosophila* and sea urchin eggs. Mazia suggested that a complete revision of our image of the cell was needed, consisting of three domains: (1) the cell surface complex – which included the plasma membrane, the outward-facing components such as receptors and coats, and a cortex underlying the

---

24 Mazia, "The Chromosome Cycle and the Centrosome Cycle," 79.
25 Mazia, "Mitosis," 393.

membrane – that encases the cell body and protects it from the external environment; (2) the cell body; and (3) a background phase in which the cell body was embedded. He imagined the cell body carrying the centrosome could be rotated relative to the surface of the cell, and this could explain how it was possible to have an axis of division that was not necessarily perpendicular to the preceding one. With its extensive highway of membranes, it could direct transport of molecules along the microtubules from the interior to the surface. The answer to how does one cell become two would be the answer to how and when in the cell cycle the cell body becomes two functional cell bodies.[26] This also implied that the cell body rather than the cell was the smallest functional unit that had all the attributes of what we call "living."

When Mazia had compared the description of the cell cycle to the historian's invention of the Renaissance, both being idealized conceptions, he certainly didn't think the term "Renaissance" would be applied to the centrosome. Yet Anastassiia Vertii and colleagues did exactly that. "The Centrosome, a Multitalented Renaissance Organelle" described the dynamic nature of the centrosome, demonstrating that it was more than simply a MTOC, just as Mazia maintained.[27] Mazia's ideas have proven to be the basis for a rich research program demonstrating that a full understanding of the cell required far more than unraveling DNA's activities. "Renaissance" literally means rebirth, and the term is apt, both in terms of the centrosome itself and the renewed interest in it. Today, transmission electron microscopy and super-resolution microscopy have been able to reveal in exquisite detail the ever-changing structure of the centrosome, literally illuminating why it seemingly disappears as well as how it changes shape as it takes on different functions. Research has shown that the "duplication of the centrosome and variations in its microtubule-nucleating capacity are driven by cell cycle dependent changes in the cytoplasmic environment."[28] As the cell enters mitosis, the centrosome transforms into a spindle pole with the addition of many microtubules. After division is completed, if the cell commits to differentiation, it moves toward the plasma membrane and can form a basal body for cilia to form. It also appears to play a role in membrane trafficking and

---

26 Mazia, "The Cell Cycle at the Cellular Level," 14.
27 Vertii, Hehnly, and Doxsey, "The Centrosome."
28 Hinchcliffe et al., "Requirement of a Centrosomal Activity," 1547.

the formation of synapses. It may act as a sensor for cellular stress, which in turn regulates whether cells progress to mitosis or not, or are even destroyed – for example, after DNA damage has occurred. Not only does the centrosome move around the cell, it seems to orchestrate the synthesis and degradation of various molecules as needed for different phases of the cell cycle by serving as the scaffolding that anchors several regulatory proteins. A Renaissance organelle indeed!

Returning to Mazia's analogy of the cell cycle as a bicycle with a growth wheel and a reproductive wheel, what is the role of the centrosome in regulating their gear ratio? To fully understand the cell cycle will require understanding how cell growth and cell division are coupled, which is undoubtedly achieved by a variety of interconnected feedback loops and checkpoint controls. As Mazia pointed out, the bipolarity of the division process and how the chromosomes segregate is determined by the centrosome duplication cycle that occurs only once each division. Although the DNA also duplicates only once each division, the two duplications do not necessarily occur at the same time. Furthermore, while the DNA is precisely duplicated, the centrosome can and does change its molecular configuration. Besides serving as a MTOC and creating the mitotic spindle, there are several reasons to think that the centrosome/microtubule system is at the heart of regulating the relationship between cell growth and cell division, again just as Mazia hypothesized. First, microtubules are everywhere! They determine the distribution and, to some extent, the dynamics of membrane compartments. The centrosome is in tight association with the nucleus and serves to anchor the microtubules. Changing the molecular arrangement of the microtubules results in a network that has an asymmetrical organization and a unique polarity. This in turn makes possible the assembly of new membranes of virtually all cellular compartments. This ability makes microtubules good candidates for monitoring the cell shape and mass increase of the cell. Second, the centrosome duplication cycle occurs across all four phases of the cell cycle. The centrosome is required for several cell cycle transitions, including G_1 to S phase, G_2 to mitosis, and metaphase to anaphase. In most cell systems, the major events of centrosome duplication anticipate major cell cycle events. For example, the duplication of the spindle pole body precedes the S phase, and the separation of the duplicated spindle pole bodies marks the beginning of mitosis. This suggests that the control of the cell cycle or the entry points of the various phases are determined by the progression of the centrosome duplication cycle. Research has demonstrated that there are some core centrosomal structures that are distinct from the elements of the MTOC that are necessary for the somatic cell cycle to progress from the G_1 phase to the S phase. Once the cell has entered the S phase, these structures are no longer needed.[29] This also supports Mazia's conjecture that the centrosome was actually more than just a MTOC. Because the doubling of the centrosome is a discontinuous process, but at the same time parallels the doubling of the chromosomes, it could be involved in maintaining the correlation between cell mass and chromosome number. Thus, the centrosome/microtubule system has the properties that could monitor time and mass increase. Finally, while great emphasis has been paid to understanding how the exact duplication and then segregation of the genetic material into two cells occurs, it must not be forgotten that it is the whole cell that divides, both the nucleus and the cytoplasm. This entails reproduction and distribution of the cytoplasmic compartments as well. While the exact duplication and distribution of these cytoplasmic components may not be as critical as it is for the genetic material, it must be reasonably accurate over the long term. How those components are distributed is crucial to normal development. It appears that the centrosome is the only component of the cytoplasm that has such a precise duplication. This is necessary to ensure the precise segregation of the chromosomes, but the centrosome could also serve as a control center for the distribution of all the other cytoplasmic compartments as well. This would certainly be more efficient for the cell than to have a separate duplication mechanism for each cytoplasmic compartment.

The correct lining up of each chromosome and then segregation is made possible by the spindle apparatus; at the same time, the flexibility of the centrosome/microtubule array also makes possible the asymmetric division of the cytoplasm. This is one of the few basic mechanisms that allows for cell differentiation and is determined by the position of the mitotic spindle in the cell, and can control the ratio of cytoplasm to nucleus during development. As the cell grows and divides, it has both spatial and topological constraints and the centrosome/microtubule array seems to underlie it all. All of this information essentially describes Mazia's

---

29 Hinchcliffe et al., "Requirement of a Centrosomal Activity," 1547.

concept of the cell body, although most researchers have not used this term. One exception is František Baluška and his colleagues, who explicitly adopted the idea of the cell body and are calling for a revision of cell theory. In a paper dedicated to Mazia, they began by pointing out that the first person to bring attention to problems with cell theory was none other than Thomas Huxley.[30] In later work, they combine the cell body concept with that of the energide, first suggested by Julius von Sachs in 1892.

Sachs, like Huxley, was well aware of certain problems with the cell theory. Recall from chapter 2 that Sachs thought the term "cell" had "originated from an unfortunate mistake." He suggested the nucleus and its surrounding area of protoplasmic influence was the smallest unit that was "living" and existed within a supra-cellular framework. He gave the name "energide" to this concept. He had observed in **coenocytic** (containing many nuclei within one cell) alga that a nucleus always organized the area of cytoplasmic space that surrounded it, regardless of whether or not it was enclosed by a cell membrane. The cell periphery or membrane was a secondary structure that was generated by the energide, and the cell might contain one or more energides. Agreeing with Huxley, he recognized that the processes inside the cell were what made the cell "alive." With the state of microscopy at the time, Sachs could not have seen the system of microtubules, but it is indeed impressive that more than 100 years ago he proposed a concept that has been validated by current research. Even by the early twentieth century, it had been observed that the early stages of *Drosophila* development occur in syncytia (tissue that contains hundreds of nuclei in a common cytoplasm). Modern-day research has shown that the nuclei are not active until the twelfth nuclear division. The cytoplasm differs from place to place, and the nuclei and their surrounding microtubule-rich cytoplasm are called energides. Baluška now uses the term "neo-energide" rather than cell body. This is not just because Sachs had proposed essentially the same idea much earlier, but rather because "the term energide better invokes the unique properties of this universal unit of supra-cellular living matter endowed with the vital energy."[31] Also, while Mazia thought "something was missing in our concept of the cell," it is not clear that he would have gone as far as claiming that the cell body should replace the cell as the smallest independent unit that has the quality of "life," as Baluška and his colleagues are advocating.

---

30 Baluška, Volkmann, and Barlow, "Eukaryotic Cells and Their Cell Bodies," 9–32.
31 Baluška, Volkmann, and Barlow, "Cell-Cell Channels and Their Implications," 1.

REVISITING PROBLEMS WITH CELL THEORY

When Hooke first discovered the cell by examining cork, he certainly had no idea that it would become the basis for a theory that defined the fundamental unit of life. His concept had little in common with the cell of cell theory, but his description, with its emphasis on the wall as a partition or a boundary that separated one cell from another, has dominated how we think about cells and organisms to the present day. Many researchers today would agree with Sachs that the adoption of the term "cell" had been regrettable in the quest to understand life. It is somewhat ironic that Hooke's discovery was made in a plant, because plants have always presented difficulties for cell theory for a variety of reasons. Sachs was not the only botanist who found the idea of the cell problematic. Eduard Strasburger, who is often referred to as the founder of modern plant cell biology, recognized that certain aspects of plants were exceptions to cell theory. He made many contributions to understanding the cell and cell division. Most notably he character-ized in detail the first three stages of mitosis and named them prophase, metaphase, and anaphase, as well as describing **cytokinesis** (the cyto-plasmic division of the cell at the end of mitosis and meiosis). He also coined the word "protoplasm," which later was applied to animal cells as well. He observed cytoplasmic streaming and showed that just as the cell only arises from preexisting cells, the same was true for the nucleus. Like Huxley, Strasburger also observed that virtually all plant cells were connected through cell-to-cell channels, today referred to as **plasmodesmata**. They formed across the wall as the cell was dividing and also occurred in already established walls. This means that potentially all the nuclei can be in direct contact with every cell via the plasmodes-mata, creating an integrated superhighway that traverses the entire plant. In addition, many species of plants as well as fungi have cells that are multinucleate, forming coenocytes. While mitosis seems to occur almost automatically in diverse organisms, cell division does not neces-sarily follow. For example, in most plants the endosperm tissue is filled by actively dividing nuclei, but the cell wall is not formed until much later, after the coenocyte has formed. A similar process had already been described in various species of algae by Strasburger. Coenocytes are also characteristic of early stages of animal development and the formation of spores in fungi. In *Drosophila* embryos, for instance, the nuclei divide

13 times without cell division. Only at cycle 13 do cell membranes form. Mammalian sperm are derived from stem cells that retain connections to each other after they divide. These observations had been made long ago, yet according to cell theory, the cells of plants and animals were considered to be entirely autonomous and separated from their surroundings by the plasma membrane. We now know that a supra-cellular organization exists in many animal species as well and allows for the exchange of organelles as well as nutrients. In animals, bridges between cells can be generated *de novo*. Thus, plants, animals, fungi, and algae have independently evolved both multicellularity and **supra-cellularity**.

This supra-cellularity seems to have been particularly useful for sessile plants to adapt to life on land and evolve in hostile environments. The continuity of cellular units allows potentially unrestricted exchange of information throughout the plant body, the informational signals being used to rapidly coordinate genome transcription that can either neutralize or take advantage of environmental challenges.[32] Baluška, in fact, argues that only animals can be considered truly multicelled and that it is more appropriate to describe higher plants as consisting of communicating cytoplasms. A truly amazing example of intercellular communication in the plant world is Pando (Latin for "I spread"), often referred to as the Trembling Giant. Discovered in 1968 in south-central Utah along the western edge of the Colorado Plateau, Pando looks as if it is a large stand of quaking aspen, but it is actually a colony of a single male organism of *Populus tremuloides*, the individual trees interconnected by one massive root system. It is estimated to weigh about 6 million kilograms and is the heaviest known organism. The root system is thought to be about 80,000 years old, which also makes it one of the oldest living organisms.

At the other end of the spectrum of life are **prokaryotic** organisms, cells that have no organelles, including no nucleus, and thus they also did not fit into the scheme of the typical cell. Much smaller than **eukaryotic** cells, their relatively simple structure suggested that they were quite ancient and possibly had a different origin than that of plants and animals. Were they an exception to Virchow's dictum that all cells arise from preexisting cells? Were they even cells? It was not until the first part of the twentieth century that it was definitely accepted that yes, bacteria are cells, and yes, they do arise from preexisting cells.

---

32 Baluška, Volkmann, and Barlow, "Eukaryotic Cells and Their Cell Bodies," 9–32.

However, as our knowledge about the fine structure of both eukaryotic and prokaryotic cells increased, another problem for cell theory arose, and that problem revolved around the role symbiosis played in the evolution of the cell and complex multicelled organisms.

Symbiosis refers to the close association between two organisms of different species. One of the first known examples of symbiosis was the recognition that a **lichen** was not a single organism, but consisted of two interdependent organisms of different species living in close contact, each mutually benefiting the other. In 1867 the Swiss botanist Simon Schwendener proposed that lichens were a combination of a fungus and an alga or cyanobacterium, citing as evidence his extensive microscopic investigations of the anatomy and development of algae, fungi, and lichens. However, many leading lichenologists disagreed. The impact of the relatively new cell theory was already dominating their thinking, and they claimed that all living organisms were autonomous. Gradually, evidence accumulated that revealed the true nature of the relationship between the alga and the fungus. Lacking chlorophyll, a fungus cannot photosynthesize. It benefits the alga, however, by providing a substrate and aiding in the absorption of nutrients and water that allow the alga to make food for both of them. Fascinated by the partnership of the two organisms that made up the lichen, Konstantin Mereschkowsky (1855–1921) became a leading expert on them. He thought that not only lichens were the result of a symbiotic relationship. He proposed a theory of **symbiogenesis**, arguing that more complex large cells evolved from a symbiotic relationship of smaller ones. However, his theory was essentially ignored for over half a century. One possible reason his ideas were disregarded was because he rejected Darwinian natural selection, arguing that it couldn't account for the origin of biological novelty. Instead, he claimed that the acquisition of microbes in cells was the major driver of evolution.

While Mazia was claiming that the centrosome and its elaborate system of microtubules was critical to our understanding of the cell, Lynn Margulis (1938–2011) also argued that we had to fundamentally change how we thought about cells. Mazia's ideas had implications for cell theory, while Margulis's ideas suggested we needed to also revisit neo-Darwinian theory. Beginning in the 1960s, she further developed and popularized Mereschkowsky's ideas but, like Mereschkowsky, she also was marginalized for many years. She proposed that organelles such as the chloroplast and mitochondria were once free living cells.[33] With the discovery that they contained DNA that was distinct from nuclear DNA, her idea gained more credibility. However, it wasn't until the advent of detailed DNA sequencing that her ideas were really fully accepted. Again, we see that the development of new technology played a critical role in advancing our knowledge of the cell. At the same time we see how the first two tenets of cell theory also hindered the accept-ance of her ideas. Some DNA-sequencing data indicate that the nucleus was also the product of a fusion event. It suggests that an ancient pro-to-cell engulfed several bacteria and **archaea**.[34] However, not everyone accepts this interpretation. A staunch evolutionist, Margulis called her-self a Darwinian, but not a neo-Darwinian. Neo-Darwinism has primar-ily emphasized the competitive aspect of natural selection as the main driver of evolution. She argued instead that cooperation and specifically symbiotic partnerships have been central to the expansion and enor-mous diversity of life. The mainstream evolutionary biology community has strongly resisted such a view. This in turn has influenced how biolo-gists think about the role of the cell in multicelled organisms.[35]

Because cells are communicating with one another and exchanging nutrients within multicelled organisms, and are therefore not independ-ent, one might question whether the cell is the smallest independent unit that has all the characteristics of life. One could even ask what is a cell? The various organelles are remnants of cells within cells, and many bac-teria also exist in cells. In addition to mitochondria and chloroplasts, a variety of other tiny organelles are found in cells. Some contain DNA, some do not, but they all can reproduce themselves. All of them are probably products of cell merging that became simpler over time as they entered into a symbiotic relationship with the larger cell. Cells with a sin-gle nucleus may range in size from a microscopic protist to the egg of an ostrich. Coenocytes in some species of marine algae can be several meters in length, each nucleus organizing its own set of microtubules and cyto-plasmic areas. A particularly dramatic example of a gigantic cell is what happens to the placenta in the developing embryo of mammals. The sur-face becomes highly vascularized as the villi invade the uterus to establish circulation between the embryo and the mother. It is multinucleated,

---

33 Margulis, "Serial Endosymbiotic Theory," 172–74.
34 Sapp, *The New Foundations of Evolution*, 313.
35 Baluška and Lyons, "Symbiotic Origin of Eukaryotic Nucleus," 48–54.

and the surface area can be as large as ten square meters! In light of this enormous variance, might it be better to find an entity that is capable of growing and dividing and is much more uniform in its size across all the different kingdoms of life? Mazia's cell body seems to be just such a unit.

IN FAVOR OF THE CELL BODY

The cell body is capable of self-organization and reproduction and is responsive to many different external stimuli. While the cell body only reproduces once per cell cycle, it invariably precedes cytokinesis, and the flexibility of the microtubule/centrosome complex allows it to divide independently of the cell. Evidence suggests that the **endosymbiotic** acquisition of the nucleus happened before that of other organelles and might even be the first example of cooperation at a cellular level. This would explain that while the timing of cell division and mitosis are tightly coordinated, they nevertheless remain somewhat independent of each other, reflecting the origin of the nucleus.[36] Strasburger's ideas foreshadowed these findings. In his detailed studies of cell division of multinucleate cells, he showed the various ways mitosis and cytokinesis were coordinated. These investigations, however, made him realize that the two were "uncoupled" – that they were independent of each other. Mazia had always emphasized in his investigations that the cell cycle was more than simply what was being observed in mitosis. In characterizing the mitotic apparatus, he became increasingly convinced that there was something smaller than the cell, but included more than the nucleus, that was critical to understanding the cell cycle. This is what led him to his concept of the cell body. The cell body concept, even if not explicitly cited as such, has been useful not only in investigating how and when cells divide, but has also provided additional insight into how cells develop into complex multicelled organisms.

A variety of seemingly unrelated and also poorly understood phenomena benefit from making use of the cell body concept. The amount of DNA varies enormously from species to species, but it does not correlate to the organism's complexity. For example, the liverwort has 18 times as much DNA as humans, and a species of salamander has 26 times as much. An amoeba has 200 times as much DNA as a human! This is referred to

---

36 Baluška et al., "Strasburger's Legacy," 1151–62.

as the C-value enigma. We now know that much of the DNA is noncoding, and thus it doesn't correlate to the actual number of genes, but this raised another set of problems. What is all this noncoding DNA doing? Why do some species have so many genes, while others have so few? What factors contribute to the evolution of genome size, and how and why are these noncoding regions gained and lost? It appears that the noncoding DNA directly affects cell size. However, there was no good explanation for this. Remember, Mazia had suggested that the centrosome was the organ in which the expression of genes might be regulated, and research has borne this out. DNA-binding proteins that are stored within the nucleus regulate tubulin polymerization, which explains the dynamic nature of microtubules in the cytoplasm.[37] This in turn determines access to other proteins in the cytoplasm that regulate the binding of those proteins to the DNA in the nucleus irrespective of coding regions. Initially, noncoding regions of the genome were referred to as junk DNA. Various scenarios were suggested to explain it. With natural selection still the dominant paradigm in biological explanation, it was argued that if these fragments of DNA weren't harmful, they wouldn't be selected against and they would just accumulate. This, however, really isn't true. Eventually all this "junk" DNA would become a great metabolic burden for cells to be constantly duplicating. However, from the point of view of the cell body, there really is no difference between coding and noncoding DNA. Both would interact either directly or indirectly with the sequestered nuclear proteins. Rather than junk, the cell body concept implies that the noncoding DNA is a central player that links the DNA-based nuclear chromatin to the tubulin-based cytoskeleton. This makes possible additional information to be encoded within DNA sequences that are then transcribed to the RNA, which influences gene expression. In addition, recent data suggest that much of the "junk" codes for regulatory RNAs.

A great deal of research, while not explicitly adopting the term "cell body," has essentially provided evidence for the concept and at the same time offers explanations for many unanswered questions. Cell bodies are capable of sensing electric and magnetic fields, which can be used as cues to orient the poles to determine on which plane mitotic division takes place. In some cells, how the microtubules orient themselves is also

---

37 Baluška, Volkmann, and Barlow, "Nuclear Components with Microtubule Organizing Properties," 91–135.

dependent on gravitational fields. They can also sense a change in temperature. It has been suggested that the microtubules act like nerves as they can perceive and transmit light. By making use of information from the cell periphery apparatus, the cell body has the ability to sense the various different properties of the physical environment. This contributes to its ability to regulate cell division. The centrosome/microtubule complex (which is essentially a particular form of the cell body) makes possible the accurate duplication and segregation of the chromosomes. In addition, it can bestow exploratory and sensory properties on the generally passive DNA-storing nucleus.[38] This property has profound implications for morphogenesis, that is, how an egg becomes a tiger, a redwood, or a human. However, the lion's share of scientific efforts has been and continues to be focused on DNA, rather than looking at the whole cell.

The emphasis is placed on DNA for many reasons. But certainly one reason was the formulation of what Francis Crick called the Central Dogma. More than half a century ago, he argued that DNA makes RNA makes Protein, and the flow of information is one way only: DNA → RNA → PROTEIN. Once the information passes from DNA to protein, it cannot get out again. The use of the word "dogma" was controversial. As Crick wrote in his memoir, *What Mad Pursuit,* the use of the term dogma had caused him a lot of grief, but it is also interesting to see why he picked the word:

> As it turned out, the use of the word dogma caused almost more trouble than it was worth. Many years later Jacques Monod pointed out to me that I did not appear to understand the correct use of the word dogma, which is a belief that cannot be doubted. I did apprehend this in a vague sort of way, but since I thought that all religious beliefs were without foundation, I used the word the way I myself thought about it, not as most of the world does, and simply applied it to a grand hypothesis that, however plausible, had little direct experimental support.[39]

And Crick was right on both counts. The central dogma was a very powerful idea that was extremely useful in working out the details of protein

---

38 See Baluška, Volkmann, and Barlow, "Eukaryotic Cells and Their Cell Bodies," 9–32, for a summary of these findings.
39 Crick, *What Mad Pursuit,* 109.

synthesis. It was plausible, but initially there was little evidence for it. A "grand hypothesis," it provided an organizing principle to generate the necessary experimental evidence to prove its validity. The central dogma summed up what was eventually demonstrated in great detail: not only how the genetic information is passed on from one generation to the next, but also how that information actually is first transcribed and then translated into the specific molecules that contribute to particular traits. Whether one's blood type was A or B, whether one had curly hair or straight, whether Mendel's pea plants were going to produce wrinkled or round seeds, or whether the family cat was blue-eyed or yellow-eyed – all could now be explained in terms of the central dogma. Yet we have known for some time that there are many exceptions, and that the information flow is in two directions. Retroviruses such as HIV consist of RNA that is transcribed into new DNA molecules. Epigenetics is a burgeoning field that examines how processes such as methylation of the DNA affect gene expression. Furthermore, these changes in gene expression can be inherited. One could argue that these examples do not falsify the central dogma. When Crick used the word "information," he meant the specific sequence of the bases in the DNA, and in none of these cases is the sequence actually changed. However, the central dogma is being challenged in a more profound way. It has outlived its usefulness in that it gives total primacy to the information contained in the DNA sequestered in the nucleus.

The answer to development does not lie solely in the "information" that is coded in the DNA. Returning to the controversy at the turn of the nineteenth century over what controls development, the nucleus or the cytoplasm, evidence now shows that the complete answer lies in between, in the interaction between the two. In addition, many botanists working in cell biology think that the whole organism rather than the individual cell should be considered the fundamental unit of plant development. Regardless of one's views on these controversies, the cell body concept provides a different way of viewing the cell, and in doing so helps resolve many limitations of cell theory as it was originally formulated. In spite of a considerable body of research that continues to provide more and more evidence in favor of Mazia's idea about the centrosome/cell body, beyond Baluška's group virtually no one has adopted either the term "cell body" or the neo-energide concept.

RESISTANCE

Why has there been such resistance to adopting the cell body/neo-energide concept? One answer is that Baluška's group is not just providing evidence for it, but they are also suggesting that the cell body/neo-energide is the fundamental unit of life. In doing so they are challenging key aspects of cell theory. An essential addition to cell theory was Virchow's dictum *Omnis cellula e cellula*. In 1882, Flemming modified it by claiming **Omnis nucleus e nucleo**. Weiss did not disagree, but also maintained that *Omnis organisatio ex organisatione*. And now, Baluška and his colleagues are arguing for a further modification to cell theory. Cells can only arise from other cells because the energide reproduces itself first: **Omnis energide e energide** (energides only come from energides). This represents a direct threat to the primacy of the cell. Needless to say, such a fundamental idea in biology is going to be very difficult to displace. The plausibility of the neo-energide as the smallest independent unit of life also depends on accepting an endosymbiotic origin for the nucleus. In spite of considerable evidence in support of this idea, it remains controversial. Baluška's group proposes that a hypothetical actin-based host cell was invaded by a hypothetical tubulin-based guest cell. The nucleus is a remnant of this first endosymbiont and has maintained its autonomy. The cytoplasm and the cell periphery complex are the remnants of the host cell, but its activities are now tightly controlled by the nucleus (see figure 7.4).[40] While some cell biologists accept this endosymbiotic scenario for the origin of the nucleus, others claim that it was generated from a single prokaryote through the invagination of its plasma membrane. Another group thinks that viruses were primarily responsible for the creation of the nucleus.[41]

Until consensus is reached regarding the endosymbiotic origin of the nucleus, the cell body/neo-energide concept will probably not be widely adopted. Nevertheless, there is much research in other disciplines that supports Huxley's criticisms about cell theory. As earlier researchers with much less empirical information recognized, heredity and development were only complementary aspects of the more fundamental question of generation – or, in Mazia's words, how one cell becomes two. Revisiting cell theory has profound implications for evolutionary theory as well,

---

40 Baluška and Lyons, "Symbiotic Origin of Eukaryotic Nucleus," 2018.
41 See Nicholson, "Biological Atomism and Cell Theory," 209.

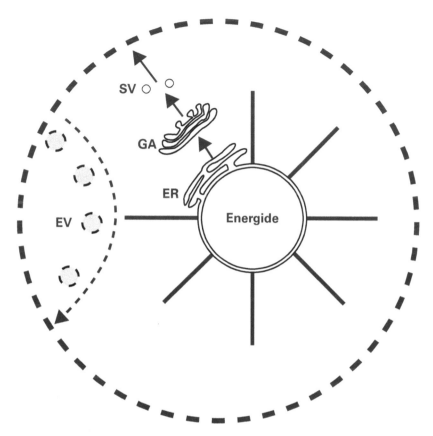

7.4 Energide (cell body), shown in continuous lines, settled within a highly reduced guest cell, shown in dotted lines, which serves predominantly as a shelter. ER: endoplasmic reticulum, GA: Golgi apparatus, SV: secretory vesicles, EV: endocytic vesicles.

particularly the role of symbiosis in the evolution of eukaryotic cells and multicellularity, which in turn has resulted in the incredible **biodiversity** that is found on planet earth. To return to Huxley's primary interest as a morphologist, natural selection did not play a large role for him in trying to understand the laws that guided development. Yet just as heredity and development need to be brought back together to understand generation, development and evolution need to be integrated to answer Huxley's question: how does form come to be? The next chapter explores how the relatively new discipline, evolutionary developmental biology (evo-devo), and the even newer discipline, ecological developmental biology (eco-devo), are addressing concerns that Huxley articulated more than 150 years ago.

CHAPTER EIGHT

How Does a Chicken Become an Egg? Evo-Devo and Eco-Devo

Evolution is the control of development by the environment.
Leigh Van Valen, "Festschrift," 1973

AND HOW DOES AN EGG BECOME A CHICKEN?

Darwin's theory of evolution argued that organisms change through time and that the primary mechanism was natural selection. Natural selection, as we have seen, was not well accepted until the twentieth century. However, descent from a common ancestor was quickly adopted, primarily because of the impressive evidence from embryology. Indeed, Darwin recognized it was the strongest evidence in favor of his theory. Thus it is somewhat surprising that later evolutionary biologists were not more interested in trying to incorporate development into their research agendas. Not until the advent of evolutionary developmental biology in the 1980s did the kinds of questions that interested Huxley begin to garner significant attention. Likewise, Huxley's most fundamental criticism of cell theory, that the cell cannot be considered an independent unit, is finding resonance with an even newer discipline, ecological developmental biology. To understand how a cell becomes an organism, it is important to note that not only is the cell communicating with other cells within the organism, and the organism is informing the cell, but at the same time the organism is also being informed and shaped by the environment. Organisms develop and evolve within ecosystems, each influencing the other. This research also provides evidence in support of a revision of cell theory. In addition, ecological developmental biology (eco-devo) also suggests that the highly reductive program of molecular biology, which has been so spectacularly successful, is inadequate to fully answer the problem of form. This organismal view of life reappears in every generation of biologists but, as this history shows, has always been a minority viewpoint. Yet today more and more research is being informed by a perspective that makes use of both bottom-up and top-down approaches, the embodiment of organicism. While the fruit fly was the early organism of choice for genetic studies and the sea urchin for development, many different kinds of organisms are contributing to the research in eco-devo and evolutionary developmental biology (evo-devo). Slime molds, in particular, are providing new insights and highlight all that we still don't understand about both development and evolution.

SUBLIME SLIME MOLDS

"Look up, it's a bird. It's a plane. It's Superman" began the iconic television series that brought the comic book superhero to life. A very different show might well begin "Look down, it's a plant. It's a fungus. It's an animal. No, it's a **slime mold**." This slimy (hence the name) blob-like creature seems like an unlikely superhero in the world of life, but the more we learn about it, the more amazing are its characteristics. For our story, it epitomizes in many ways the limitations of cell theory in trying to understand how a single cell grows and becomes an organism, communicating and functioning within an ecosystem. As researchers continue to unravel its secrets, the slime mold is providing insight into answering biology's most fundamental questions.

For a long time slime molds were thought to be fungi since they can't photosynthesize, they feed on decaying matter, and many species at least superficially look quite similar to fungi. However, that was always a problematic classification because, unlike fungi, slime molds can move. Yet some are unicellular, which means they are not classified as animals. While some are microscopic, they are different from bacteria because they have true nuclei. Like many other oddball organisms that defy easy classification, they have found a home in the kingdom Protista. Slime molds are divided into two major groups: the cellular slime molds, which are truly multicellular with different cell types, and plasmodial slime molds, which are not. Both groups vary enormously in how they look. Some are microscopic, some grow very large, and they come in all kinds of shapes and colors. More than 900 species have been identified. Initially cellular slime molds were primarily used to study cell differentiation, but the plasmodial slime molds are turning out to be equally important. In a myriad of different ways, these fascinating creatures exemplify all we still don't understand in regard to cell differentiation, development, and evolution. They are even making us rethink how we define intelligence and how it evolves.

One is most likely to encounter slime molds in the woods clinging on to rotting logs. The cellular slime mold *Dictyostelium discoides* spends much of its life cycle as a haploid amoeba, creeping along as it digests spores, bacteria, and other microbes. It apparently can remain like this indefinitely, as long as there is enough food. However, once it exhausts its food supply, the amoeba secretes a chemical message, cyclic adenosine monophosphate (cAMP), which induces them to migrate to the source of the chemical and bind to their fellow amoebas. Individual amoebas swarm together to form streams of amoebas that join together, forming a giant slug containing thousands of cells, each retaining its individual cell wall. Inside the slug, a small number of the cells take on the role of the white cells of our immune system as they move through the slug in search of pathogenic bacteria that they devour. These then drop off the slug, leaving the healthy cells intact. The slug creeps along toward the light and, if conditions are right, something quite remarkable happens. It stops, upends itself, and turns into a true multicelled organism as the individual cells differentiate. The bottom cells serve to anchor the whole organism, while the middle cells start to grow a stiff stalk, turning their insides into bundles of cellulose. Cells at the tip grow into a fruiting body consisting of the spore cap and thousands of spores. Wind or rain will knock off the cap, dispersing the spores. Under suitable conditions the spores germinate, turning into amoebas and the cycle continues. Meanwhile, the cells that made up the stalk die. Many species of plants such as wheat produce stalks that contain seeds that are dispersed by the wind, after which the stalk withers and dies. However, what is different is the wheat cells have spent their entire life as part of a multicellular plant. In the case of the slime mold, the individual ameboid cells probably originated from many different fruiting bodies that came together, and the ones that eventually form the stalk die, essentially sacrificing themselves for others. This suggests a kind of altruism, where the individual sacrifices itself for the good of the group, something that was thought to only exist in more highly developed organisms.

The evolution of altruism has been a topic of contentious debate among evolutionary biologists ever since Darwin. Some argue that true altruism does not exist, that it results from kin selection, is ultimately selfish, and is just a way of getting one's genes into the next generation. Without getting into the details of that debate, the slime mold's behavior is at the very least a type of cooperation, and shows that cooperation evolved very early in the history of life. DNA sequencing suggests that cellular slime molds evolved around 600 million years ago, and the common ancestor of all slime molds may have evolved over a billion years ago. It was one of the earliest pioneers on land, arriving hundreds of millions of years before the first plants and animals. Cooperation, then, has been an important strategy in, to use Darwin's words, "the struggle for existence."[1] Just like the kinds of symbiosis that were discussed in the previous chapter, the slime mold contradicts the prevailing "nature red in tooth and claw" characterization of evolution. Instead, "survival of the fittest" may mean survival of the cooperators.

Dictyostelium discoides is not the only slime mold that has researchers scratching their heads in amazement. *Physarum polycephalum*, which means "many-headed slime mold," seems to have the ability to make decisions and exhibit intelligence. Like its cousin, it also exists in an ameboid state with individuals coming together in times of food shortage. However, unlike *Dictyostelium*, it becomes one giant cell containing thousands of nuclei, transforming itself into a pulsating blob, oozing along in search of food. In 2000, Japanese researchers Toshiyuki Nakagaki and colleagues brought it into the lab to study how it moved. They cut one slime mold into lots of little pieces and scattered them in a plastic maze. The pieces grew, finding each other as they traversed

---

1 Darwin, *Origin of Species*, chapter 3.

the entire maze. The researchers then put blocks of agar with nutrients at either end of the maze. After four hours, the slime mold had retracted all its branches that led to the dead ends. In another version of the experiment, the slime mold grew only along the shortest path between the two pieces of food. They are extremely efficient at doing this. Canadian researchers placed rolled oats (slime molds really like oats!) on the major population centers of a map of Canada and the slime mold was placed on Toronto. It worked its way across the entire map, sprouting its tendrils that mimicked the Canadian highway system. This experiment has been repeated using maps of the United States, Tokyo, and London. The slime mold moves in a pattern that resembles the highway system of the United States, the railroad system of Tokyo, and the London underground. Quite impressive for an organism that doesn't even have a neuron, much less a brain! The implications of this are quite profound.[2] Andrew Adamatzky and colleagues have been fascinated by slime molds' ability to mimic highway systems. In 2010 they placed a slime mold in the middle of a map of Spain and Portugal, with pieces of food on the largest cities. The slime mold grew a network of tentacles that was nearly identical to the actual highway system on the Iberian Peninsula. Thus Adamatzky recommends that, in developing countries where highway systems are being built from scratch, the designers should follow the routes that the slime molds make on a map of the country that has food placed on the population centers. In another study, he used them to simulate a nuclear disaster. First, he let the slime mold grow a network of highways for Canada. He then placed a crystal of sea salt, which repels slime molds, on the map at Lake Huron in Ontario, where the Bruce nuclear power plant is located. The slime mold abandoned its tendrils near the salt and grew a new highway pattern that efficiently rerouted food across Canada. "Reactions to spreading contamination may shed some light on what would happen if real disasters occur."[3]

The interest in slime molds now extends far beyond biologists, as researchers try to discover the biological principles that underlay its computational rules and then apply them to the fields of electronics, programming, and robotics. Some truly fascinating work is being

2 Jabr, "How Brainless Slime Molds Redefine Intelligence."
3 Adamatzky and Jones, "Road Planning with Slime Mold."

done by the ecoLogic Studio and the Urban Morphogenesis Lab. They are using the principles of biological self-organization as embodied in *Physarum polycephalum* in a conceptual model of what they call the world's Urbansphere, and suggest how it can sustainably co-evolve with the Biosphere. Their article "Cities as Biological Computers" envisions "a future where manmade infrastructures and nonhuman biological systems will constitute part of a single biotechnological whole." The authors created a physarum machine using the slime mold and have shown the feasibility of its use to a specific case: the area of eastern Arizona ecologically depleted as a result of mining. They suggest that future research will make it possible to harness the computational power that is latent in microorganisms such as the slime mold and apply it to solve complex real world problems.[4] The artist Heather Barnett has formed the Slime Mould Collective, "an international network of/for intelligent organisms" that has attracted people from all walks of life as they share their thoughts, their art, and their research on slime molds.[5] The slime mold serves as a symbol to explore ideas and engage people in a sense of community – once again fairly impressive for a blob of oozing protoplasm. These new areas of study with *Physarum* have also served as an impetus for further studies with *Dictyostelium* to understand the process of cell differentiation, the critical aspect of understanding development. Research on slime molds in many ways illustrates the importance of integrating heredity, development, and evolution in our quest to understand life. Just as the split between development and heredity had important consequences for understanding development, the lack of incorporating development into the modern synthesis also had important consequences for evolution's explanatory power in understanding the history of life. Once again, Huxley's writings serve as a jumping-off point for the need to bring the findings from embryology into evolutionary theory.

EXPANDING THE MODERN SYNTHESIS

In 1878, Huxley wrote a long essay, "Evolution in Biology," for the *Encyclopedia Britannica*, and it remained the entry for Evolution until

---

4　Pasquero and Poletto, "Cities as Biological Computers," 10–19.
5　Barnett, The Slime Mould Collective.

the early 1900s. Huxley began his article by pointing out, as was discussed in chapter 4, that the original meaning of evolution had always been associated with development, although researchers differed in how they thought development occurred. For Charles Bonnet evolution simply meant the expansion of what was invisible to the visible, and that all living beings developed from preexisting germs. In his earlier writings he seemed to have subscribed to the hypothesis of *emboîtement*, but in later writings he admitted that the germ wasn't necessarily an actual miniature of the organism. It could be merely an original "preformation" that was capable of producing the organism. Defined this way, however, the "evolution" of such a germ would not be distinguishable from the meaning of epigenesis. Many of the earlier ideas of how development occurred were found to be untenable, but by the 1800s the terms "development," "*entwickelung*," and "*evolutio*" still referred to the processes by which embryogenesis occurred. As Huxley wrote, "[Evolution] is in fact at present employed in biology as a general name for the steps by which any living being has acquired the morphological and the physiological characteristics which distinguish it."[6] Thus, in the first part of Huxley's article the word "evolution" did not refer to species change through time, but rather to how an individual form changes through time as development proceeds. Huxley drew on the ideas of Descartes and elaborated by Herbert Spencer, giving a much more general account of the word "evolution." All objects of the physical universe, whether living or not, have "originated by a process of evolution, due to the continuous operation of purely physical causes, out of relatively formless matter."[7] Biological evolution was just one aspect of a more general law of evolution that referred to a spontaneous ordering or self-ordering, and in which the germ or cell had a particular chemical composition that exhibited the properties of what we call life. Spencer had defined evolution as the "transformation of the homogeneous into the heterogeneous, the indefinite into the definite or the transformation of the incoherent into the coherent," which was actually very close to von Baer's laws that explained how development proceeded. This is undoubtedly one reason why Huxley preferred Spencer's definition of evolution over Darwin's.

---

6 Huxley, "Evolution in Biology," 282.
7 Huxley, "Evolution in Biology," 291.

Not only is development the key to understanding how a single cell becomes a complex, multi-billion-celled organism, but it is also thoroughly connected to evolution, since it is through changes in the embryo that changes in form arise. The question that most interested morphologists such as Huxley was how exactly does form emerge? Natural selection only provided a partial answer. As E.S. Russell (1887– 1954) in his classic work *Form and Function* documented, one of the most fundamental debates in nineteenth-century biology was over what was primary: form or function.[8] Although many developmentalists argued against Darwin's theory, they were not just misunderstanding the theory or tied to backward ideas. Rather, they raised issues that remained problematic throughout the twentieth century. Von Baer maintained that "usefulness" was an incomplete criterion for evolutionary change because, as he pointed out, there would have been many times in evolutionary history that having four legs and two wings would have been advantageous, yet such a form never evolved. He believed internal factors limited the types of variability produced that natural selection could act on. For von Baer, these constraints meant that natural selection could not account for the origin of his four distinct types of organisms from a single primeval germ or cell.

Paleontologists had always been sympathetic to the anatomists' concept of the *bauplan*, the idea that a general blueprint or plan exists that described the common and homologous features of a particular group or taxa. This was because the pattern of the fossil record suggested that evolution has involved the elaboration of a relatively small number of combinations of major body parts. Paleontologists refer to the possible morphologies that organisms can occupy as morphospace. The fossil record documents that not all theoretically available morphospace is filled. Just as a creature with four legs and two wings never evolved, neither did a creature with wheeled appendages because nerves or a blood supply can't connect to a rotating organ. Nevertheless, having wheels would have been extremely useful. Several explanations have been offered. Functionalists argue that the missing forms have been selected against. Historicists claim that a particular lineage is constrained historically and hasn't diffused sufficiently to fill the space. Contingency and chance, including the effect of mass extinctions, play a large role

---

8 Russell, *Form and Function.*

in their interpretation of the fossil record. However, another explanation suggests that certain forms have never evolved because of developmental constraints that limit the possibilities, as von Baer argued. Not just any form can evolve, no matter how adaptive it might be, and a complete explanation for the pattern of the fossil record must draw on research from developmental biologists. These ideas finally began to get a serious hearing in the 1980s as an outgrowth of debates over punctuated equilibrium.

In 1972 Niles Eldredge and Stephen Gould challenged the Darwinian paradigm of slow, gradual change or phyletic gradualism with their theory of punctuated equilibrium.[9] They argued that the fossil record documented a pattern of periods of rapid diversification interspersed with periods of relatively little change or stasis. Darwin argued that the gaps in the fossil record were due to its incompleteness, and such a claim was reasonable at the time. Paleontology was a brand new discipline, and most of the earth had not been explored in search of fossils. Since Darwin's time, sampling of the fossil record has become far more extensive and carefully done. Radiometric dating made possible the dating of fossils more accurately, and now molecular sequence data are extending our reach even further into the past. Yet explaining the pattern of the fossil record has remained problematic. Indeed, Gould and Eldredge began their 1977 paper on punctuated equilibrium with Huxley's caution to Darwin on the eve of the publication of *The Origin*: "You have loaded yourself with an unnecessary difficulty in adopting *natura non facit saltum* [nature does not make jumps] so unreservedly."[10] The details of the debate have changed, but the underlying issues remain the same. In spite of millions more fossils being found, systematic gaps remain. Gould and Eldredge argued that the gaps were not due to an incomplete record, but rather were real and reflected how evolution was occurring. Punctuated equilibrium was hotly debated throughout the 1970s and 1980s, but even its harshest critics admitted that it stimulated and renewed debate among evolutionary biologists, particularly paleontologists. Today, it is acknowledged that the fossil

---

9 Eldredge and Gould, "Punctuated Equilibria: An Alternative to Phyletic Gradualism," 82–115.

10 Gould and Eldredge, "Punctuated Equilibria: The Tempo and Mode of Evolution Reconsidered," 115. For original, see Huxley, 1859, in *Life and Letters of Thomas Henry Huxley*, vol. 1, 189.

record exhibits both kinds of patterns. The revitalization of paleontology combined with the research in molecular biology resulted in what is now one of the most exciting areas in evolution studies: evolutionary developmental biology or "evo-devo," which seeks to obtain a full integration of embryology with evolution.

EVOLUTIONARY DEVELOPMENTAL BIOLOGY

The diversity of life that we observe today is a product of both **phylogeny** (the evolutionary history of the species) and **ontogeny** (the developmental history of an individual from the embryo to an adult). Evo-devo is both a synthesis and a negotiation between the two disciplines of evolution and development. It connects genetics and evolution through development. Heritable changes in development can lead to new phenotypes that can then be selected for and spread through the population. At the same time, for all the diversity that exists, in many ways evolution is a conservative process. We now know that many genes have been conserved, literally for millions and millions of years, but have been put to many different uses. Developmental mechanisms may be the key to why some forms are able to evolve and be maintained, while others cannot, no matter how adaptive they might be.[11]

It is ironic that embryology was eclipsed with the rise of population genetics because it was always central to Darwin's thinking. For him, embryology provided the strongest evidence in favor of evolution, and the crucial concept in embryology was **homology**. Darwin wrote to G.H.K. Thwaites, "I do not look at [homology] as mere analogy. I would as soon believe that fossil shells were mere mockeries of real shells as that the same bones in the foot of a dog and wing of a bat, or the similar embryo of mammal and bird, had not a direct signification, and the signification be unity of descent or nothing."[12] Homology was first defined by the anatomist Richard Owen (1804–92). A homologue was "the same organ in different animals under every form and

---

11 Gilbert and Burian, "Development, Evolution, and Evolutionary Developmental Biology," 61–62.
12 Darwin, 21 March 1860, in *More Letters of Charles Darwin*, vol. 1, 83.

function," while an analogue was "a part or organ in one animal which has the same function as another part or organ in a different animal."[13] As important as the concept of homology is to both development and evolution, its meaning has always been somewhat ambiguous.[14] Today homology is at the heart of current research that is trying to fully integrate the findings in development with evolutionary theory. In addition, it is the basis for much important research in biomedicine. After all, testing new drugs on animals is based on the premise that their metabolism is homologous to our own.

When we say that a mammalian hormone is the same as a fish hormone, or that the human DNA sequence is the same as a sequence in a chimp or a mouse, we are making a direct statement about homology; likewise when biologists tell us that yeast, fruit flies, and frogs share many of the same genes as humans. Nevertheless, defining homology has always been a bit problematic. Darwin claimed that the "essential" similarities observed between different organisms were the result of having a common ancestry. Before Darwin, taxonomists had been classifying organisms based on similarity, although they did not know that the similarity was due to common descent. Nevertheless, how did one determine what was an essential similarity? What if animals agreed in structure, but differed in development, or the reverse – they agreed in development, but differed in their final structure? Which should take precedence? This was something Huxley struggled with as he attempted to develop a classification of the Medusa. Huxley, influenced by von Baer, gave precedence to development, and claimed organisms should be regarded as similar in development when their various organs shared similar laws of growth. Those laws were hypothetical at the time, but von Baer had provided a practical way to study them. By comparing in different species how various organs developed, such as a forelimb of a horse and that of a man, one could see they were the same or homologous. A series of forms such as the forelimb of a monkey and lemur, the forefoot of a bear, and so on existed such that one could imagine one form could conceptually be transformed into another. "*There exists a series of intermediate forms – the difference between any two of which is not greater than might be accounted for by hypothetical laws of growth pervading*

---

13 Owen, "Report on the Archetype and Homologies," 175.
14 See Hall, *Homology*.

the whole series" (emphasis in original).[15] The laws were not arbitrary because they were similar to "laws" that were already known, that is, patterns of growth that had been observed in a variety of different cases.

Although what Huxley wrote seems in hindsight to be clearly arguing for descent from a common ancestor, we have no evidence that he made the explicit connection between ontogeny and phylogeny. It remained for Darwin to do that. However, Huxley had written to Darwin while he was writing *The Origin* that the differences between organisms "result not so much of the development of new parts as of the *modification of parts already existing and common to both divergent types*."[16] This points out why he valued von Baer so highly, and also why he accepted common descent so quickly. When Huxley had first read of Darwin's theory, he even exclaimed, "How stupid to not have thought of that." Von Baer had shown that by following development, one could determine what structures should be considered homologues. Darwin, however, claimed that the reason why von Baer's method worked was because they shared a common ancestor in the past. Nevertheless, the meaning of homology was not as clear as it might seem.

After Darwin, an interesting and not uncommon transition took place: from explanation to definition. Instead of homology being evidence for evolution and a phylogenetic relationship, it became part of the definition. Homology was defined as any similarity that could be traced to ancestry. However, similar structures often arise independently; in contrast, often very different structures arise through evolutionary transformation. In determining genealogies one has to rely in large part on similarity of character. One has to make a choice as to which should be given primacy – similarity or ancestry. The problem that Huxley articulated still remained. Homology was being used to explain both the maintenance of similarity *and* the transformation of form. For a new group of researchers, the key to understanding homology would be to examine the underlying genetic mechanisms.

At the end of the nineteenth century William Bateson cataloged a variety of "monsters," aberrant forms that displayed extra, missing, or

---

15 Huxley, "Some Considerations upon the Meaning of the Terms Analogy and Affinity," quoted in Lyons, *Thomas Henry Huxley*, 70. Huxley noted in pencil on the manuscript, "Written between 1846 and 1847?"

16 Thomas Huxley to Charles Darwin 7 July 1857, "Letter no. 2119"; emphasis added.

altered parts. Those mutants in which one body part was transformed into another he called homeotic (from the Greek *homeos*, meaning same or similar). As the fruit fly became the organism of choice for genetic experiments, geneticists isolated several different homeotic flies. For example, in the *Biothorax* mutant the normally tiny hindwings looked like the much larger forewings. *Antennapedia* caused legs to appear instead of antennae on the head. These dramatic effects were under the control of a single set of genes called hox genes.

In the last 40 years biologists have worked out the details of how these homeotic mutants are created. Hox genes shape the development of body regions along the body axis of the fly. They belong to a class of genes that share a short, virtually identical sequence of DNA that geneticists named the homeobox. However, the discovery of the homeobox created havoc with the traditional distinction between homology and analogy because "homologous" genes are responsible for "**analogous**" processes. The segmentation in the body parts of flies and vertebrates was always cited as a classic example of analogy, but hox genes have been conserved through more than 500 million years of evolutionary history. The same genes were found in various insects, earthworms, frogs, mice, and even humans. The hox genes in mice were arranged in clusters, just as they were in the fly, and the order corresponded to the order of the body regions in the mouse in which they were expressed, just as they were in flies. The vertebrate eye and the insect eye were also regarded as analogous structures. However, they are both based in part on the expression of another homeobox-containing gene, Pax6. Along with cephalopod eyes, they are all descendants of a basic metazoan photoreceptor cell that was specified by the Pax6 gene.

Compared to vertebrates, flies don't have much of a heart, but they have a structure along the topside of their body that pumps fluids along the inside. They have an open circulation system, meaning the blood just bathes the tissues and is not compartmentalized. Geneticists isolated a gene they named *tinman* (after the tinman who didn't have a heart, from *The Wizard of Oz*) that was needed to make the heart. Several mammalian versions of the tinman gene have been isolated that belong to the NK2 family, which plays an important role in the development of the heart. In spite of the enormous differences between the heart and circulatory system of flies and humans, a similar gene is responsible for the formation and patterning of their hearts.

Several different families of homeobox genes have been discovered, all having dramatic effects on the formation of body parts. By binding to specific regions of DNA they turn genes either on or off in developing limbs, eyes, or hearts. They have such large effects because they either regulate large numbers of genes, act early on in development, or both. These examples along with many others have resulted in a revisiting of the meaning of homology. Homology has traditionally been associated with structure, but the discovery of these homologous genes of process is at the core of understanding development. The discovery that the same set of genes controls the formation of body regions and body parts in organisms as diverse as fruit flies, frogs, and zebras has resulted in a complete rethinking of animal history, the origins of structure, and the nature of diversity.[17]

Ever since Darwin, a persistent question has been the one that Huxley first raised: did natural selection have the power to create new species, rather than just well-marked varieties? In other words, how is novelty created? The astounding findings of the last 40 years have shown us that the creation of new forms is possible because of a few key concepts. First, evolution works by tinkering with what is already present. Wings did not spring *de novo* from a four-legged vertebrate. Second, structures can have a variety of functions. Wing structures were probably not originally used for flying, but for thermoregulation. Multifunctionality extends to the level of the gene, with the same "old" gene being used in many different ways. Third is the idea of redundancy. Simply duplicating a particular structure can generate very different forms. Any part of a multifunctional structure that is even partially redundant in function sets the stage for specialization through the division of labor. This can eventually result in the structure differing both in form and function from the original. Gene duplication has been an important source of innovation, making it possible for the same gene to be used in many different ways. Finally, modularity has been crucial to the creation of an enormous amount of diversity. Millions of antibodies can be coded for with relatively few genes because of the modularity of their structure. Through duplication, mixing, and modification of various parts, arthropods have evolved different structures with different functions

---

17 See Gilbert, Opitz, and Raff, "Resynthesizing Evolutionary and Developmental Biology," 357–72, for further discussion of integrating development with evolution.

from the same basic unit, leading to the most diverse group of animals on earth. The modular nature of digits made possible both the evolution of a long fourth finger in pterosaurs and many long fingers in bats on which to extend a wing membrane. The key to this diversity is the modular genetic logic of the switches in development, allowing one part of a structure to be modified independently of another. The fossil record documents a Cambrian "explosion," a burst of new forms never seen before. Today research in evo-devo is showing us that from a few basic building blocks, nature evolved the fundamental types of animals and body parts in the Cambrian that are found in present-day organisms. Yet from those basic forms an enormous amount of diversity continues to be generated in response to the ever-changing environment.[18]

The complete unification of evolution with development is a work in progress that is being approached on several different fronts. However, the vast majority of research is still focused on the gene, examining genetic variations in the coding and regulatory sequences of developmental regulatory genes. Phylogenetic methods have revealed that all eukaryotes share a "toolkit" of proteins that regulate developmental processes, and these methods are now routinely used to describe developmental gene networks. Evolutionary biologists are incorporating the insights from development in their understanding of the relationship between genotype and phenotype. Instead of abstract models of single loci, they are now modeling gene networks based on real organisms. This partnership holds the promise of understanding both the pattern and process of evolution at the most fundamental level.

However, several very basic questions remain. First, what kind of genetic variation plays a role in evolutionary changes in development? Second, how much genetic variation in natural populations influences development? We know an enormous amount of variation exists and also that it affects gene expression, but demonstrating a direct connection between differences in gene expression and different phenotypes is difficult. Molecular and quantitative genetics are beginning to be able to answer this question. Third, one of the most interesting questions that has been raised by the accumulating data on homology of process is how do we square the lack of morphological intermediates with the idea that many developmental genes are conserved across

---

18 See Carroll, *Endless Forms Most Beautiful.*

all phyla, even across more than one kingdom? Conversely, why can the activity of many different genes still generate similar structures? As we saw in chapter 6, Waddington attempted to answer these questions. Today, the unity of type embodied in von Baer's laws of development has support in the detailed studies of molecular genetics, but a theoretical debate rages as to the meaning of these findings. Is the conservation of developmental pathways the result of selection, or is it a property of generic systems that, in fact, is impervious to selection? Strict adaptationists argue that such conservation occurs because the optimality of these pathways has been reached and can't be improved upon. Yet some developmental biologists think that developmental pathways limit what is possible. Homology is about what is conserved in evolution. Yet how and why something has been conserved is distinct from identifying homologues. All of this research illustrates what was lost when heredity and development became separate disciplines and the necessity to bring them back together.

REVISITING THE CONCEPT OF MORPHOGENETIC FIELDS

Morphogenetic fields also played an important role in the emergence of evo-devo. Homologous developmental pathways were demonstrated in numerous embryonic processes occurring in discrete regions, the morphogenetic fields. Gilbert and his co-workers proposed that fields mediated the link between genotype and phenotype. Just as the cell and not its genome functions as the unit of organic structure and function, so the morphogenetic field rather than genes or the cells should be considered as a major unit of ontogeny whose changes bring about changes in evolution.[19]

One of the most provocative scientists exploring the connection between development and evolution was Brian Goodwin (1931–2009). His research built on the idea of morphogenetic fields that was first suggested in the work of Driesch, Boveri, and Roux. His research program emphasized the spatial order that emerged as a result of cell-cell

---

19 Gilbert, Opitz, and Raff, "Resynthesizing Evolutionary and Developmental Biology."

interactions: "The order we see in living organisms manifests itself as a result of properties that were intrinsic to complex dynamic systems." Rather than using genealogy to define homology, Goodwin suggested that the study of development would yield principles that demonstrate homology was the result of developmental dynamics, that "has the logical structure of an equivalent relationship as used in mathematics."[20] Ever since Darwin, history has been intrinsic to our definition of homology and, therefore, Goodwin's repudiation of history in trying to discover the laws of form was quite radical. In a rather polemical example, he claimed that if we answered the question of why the earth goes around the sun in an elliptical orbit, by claiming it did it last year and the year before all the way back to the beginning of the solar system, most people would not consider that a very satisfactory answer. Yet Goodwin, like Weiss, thought this was a very common type of explanation in biology. He argued that a true integration of development with evolution cannot occur until the actual generative processes that underlie stable life cycles were understood. By combining the results of such a program with reconstructing phylogenies, he thought it would be possible to tell whether the taxonomic relationship of organisms is due to history or to the dynamics of cell-cell interactions that occur during development. Nevertheless, it would in actuality be very difficult to distinguish between these two possibilities. Furthermore, is the stability observed due to developmental inertia or selection?

Goodwin devoted a significant part of his career to studying the life cycle of the alga *Acetabularia acetabulum*, in the family of Dasyclada. There is an irony to his choice of organism because *Acetabularia* was one of the organisms used to "prove" that the nucleus was the seat of inheritance. Characteristic of it and its ancestors are structures known as whorls. However, the alga does not seem to need them and sheds them soon after making them. Why does it not get rid of them altogether? The standard answer is that the trait has been inherited from an ancestor, and the making of whorls is just too deep and persistent a property of morphogenesis to be gotten rid of. They are carried along by a kind of developmental inertia. Goodwin did not deny that this was a plausible explanation, but he claimed that this points to a question rather than answers it: why are whorls made in the first place, and why

---

20 Webster and Goodwin, *Form and Transformation*, x.

are they so difficult to get rid of? It turns out that for most members of the order gameteophores, they are used in photosynthesis. Since they don't "cost" too much to make, they are maintained. This is the standard explanation in terms of natural selection and adaption. However, Goodwin suggested the answer lies elsewhere and will be found in discovering the laws that generate these structures.

Without going through the details of how *Acetabularia* makes the whorls, Goodwin and his colleagues developed equations that described the morphogenetic fields responsible for making the whorls. He did this by investigating properties of the cytoskeleton, particularly as it related to calcium gradients, and also the growth of plant cell walls. Starting from spatially uniform states, stable patterns of calcium gradients spontaneously emerged as a result of interactions from several different factors. From initially simple forms a complex morphology emerged. This emergence of form had nothing to do with genes. We can see how this research supported Mazia's belief that the structure of the cytoskeleton was playing a leading role in developmental processes. Although a great deal of variety existed in the shape of whorls between species and in a single whorl as the individual alga matured, they still represented variations on a theme that gave the entire group its taxonomic unity. Goodwin was arguing that the taxonomies that we have developed to describe the relationship of organisms to one another are not the result of trial-and-error tinkering by natural selection, but rather reflect a deep pattern of ordered relationships. In this view, natural selection plays a much smaller role, and may only filter unsuccessful morphologies generated by development.

Goodwin's program was extremely ambitious, and he admitted that it wasn't clear if such a program could succeed. Even among developmental biologists sympathetic to his perspective, Goodwin's position was regarded as extreme. Nevertheless, he has been an important thinker in regard to investigating this topic, as he tried to shift the discussion from a totally gene-centered view to a more holistic view of the organism. Like the morphologists before him, including Huxley, he did not think natural selection was particularly important to answering these basic questions regarding the generation of form. However, such a view should not be regarded as antagonistic to evolution by natural selection. Rather, Goodwin advocated that a theory of morphogenesis was needed that would not be supplemental, but rather as fundamental to biology as the principle of natural selection.

For our story focusing on cell theory, these experiments again support Huxley's assertion that the cell was not acting as an independent unit within multicelled organisms. Investigations have demonstrated unequivocally that individual cells are influenced by the organism as a whole. Neighboring cells of a particular tissue played an absolutely essential role in shaping embryonic development. Virtually everyone working in embryology continued to provide evidence for the critical role of intercellular communication in cell differentiation. However, this did not result in people criticizing the limitations of cell theory, and it would take even longer for anyone to suggest that cell theory might need to be revised. This was for a variety of reasons, but certainly one of them was that the split that divided heredity and development into two separate disciplines also shaped the kinds of questions each of them addressed.

No one disputes the power of natural selection to cause **adaptation**, that is, to shape organisms' form and function in order to maximize reproductive success. However, research on a variety of different fronts from paleontology to developmental biology brings into question whether natural selection provides a complete explanation for the history of life and the diversity of form. Huxley recognized this in his quest to understand the laws of forms and to develop a taxonomic system that reflected those laws. Neither Huxley nor Darwin knew about genes. But the full integration of ontogeny and phylogeny remains an ongoing problem. Are microevolutionary changes in gene frequency all that are needed to turn a reptile into a mammal or a fish into an amphibian? Furthermore, it is not at all apparent that either homology or diversity are fundamentally adaptive phenomena. Because of the success in the twentieth century of natural selection theory in explaining adaptive complexity, the functionalist approach has been regarded as the more powerful. However, the idea of type provided a crucial organizing principle, led to the concept of homology, and, as Darwin realized, could be used as compelling evidence for his theory of common descent. Clearly the functional approach has been a powerful one, but it is also apparent that it still has not adequately answered profound questions relating to the nature of form.

BEYOND EVO-DEVO: ECO-DEVO

Theoretical evolutionary biologist Leigh Van Valen was one of the first modern-day researchers advocating the importance of development for understanding evolution. While the importance of the environment was always recognized as the selective agent in shaping adaptation, he also realized it was crucial to development as well. Furthermore, it was key to uniting the two disciplines. As he wrote, "evolution is the control of development by the environment." Written in 1973, it was many years before this often quoted aphorism was actually translated into a vigorous research program. Today, however, a third synthesis is being called for with the new discipline eco-devo as put forth in Mary Jane West-Eberhard's *Developmental Plasticity and Evolution* and Scott Gilbert and David Epel's *Ecological Developmental Biology*. West-Eberhard's research on social insects led to her interest in phenotypic plasticity. Gilbert has long championed the importance of development in understanding biology and criticized a strictly genetic approach to understanding evolution. Epel received his PhD under Mazia and then was a long-term collaborator with Mazia at the Hopkins Marine Station, becoming a leading cell and developmental biologist in his own right. They point out that cell-to-cell communication plays a critical role in the link between the genotype and phenotype of an organism. "By itself the genetic information in the cell's nucleus cannot directly produce the many differentiated cell types in a multicellular organism: cells must interact reciprocally instructing each other as they differentiate."[21] Ecological developmental biology integrates development, evolution, and ecology; in doing so the cracks in cell theory are becoming even larger. In this synthesis, the cell is being informed by the organism and the organism is being informed by the environment. While the organism is being informed by the genome, the genome is also being informed by what to express by the environment. The cell mediates the instructions from both the genome and the environment, resulting in a flow of information that goes in both directions. To fully understanding how an egg becomes a chicken will involve a synthesis of research in development, genetics, evolution, and ecology. If evolution results

---

21 Gilbert and Epel, *Ecological Developmental Biology*, 9.

in changes in gene frequencies, those changes must be reflected in changes in the genes of the developmental program, which are selected for and transmitted between generations. However, another component must be added to this scheme. The environment not only plays a crucial role in shaping adaptation and affecting the growth and behavior of the organisms, but it does this by influencing the molecular arrangements and activities within the cell. Eco-devo looks at the real-life interactions by looking at development in its ecological context. As Epel and Gilbert write, "eco-devo is the extension of embryology to levels above the individual." A chicken egg is not going to become an elephant no matter what the environment is. Nevertheless, developmental plasticity enables the genome to generate a repertoire of possible phenotypes. However, the evolutionary origin and basis of phenotypic plasticity remains poorly understood and varies tremendously from species to species.

Dogs are the most physically diverse group of land mammals. Think about the enormous variation that has been achieved in the breeding of domestic dogs, from chihuahuas to Great Danes. Is there something about the "plasticity" of the dog genome that can generate many mutations that are small enough to be survivable, but interesting enough that they capture the fancy of dog breeders to select for? Many people suspect this is true, but it has not been proven. Yet even if it is true, the approximately 450 recognized dog breeds could not have come about "without VERY strong human selection for various traits, going back all the way to the transition from hunter-gatherer to agrarian society."[22] Contrast this with cats, which show a much more limited range in size and shape. Is there some fundamental difference between the developmental programs of cats and dogs that restricts the amount of divergence in domestic cats, and literally does not allow a leg bone to grow super long or get much shorter, or to change how big or small they can become? Furthermore, there has been remarkable conservation in the entire cat family. Unless one is an expert, one cannot tell the difference between a lion and a tiger skeleton. Remarkable similarity also exists between the big cats and the house cat, not just terms of structure but also behavior.

As was discussed in chapter 6, Waddington first described canalization, in which several alternative developmental pathways exist to

---

22 Elaine Ostander, 28 September 2017, personal communication.

produce a particular structure, but they still gave the same end result. He suggested that natural selection created these pathways as a buffer against external perturbations, such as unfavorable temperatures, and internal ones, such as mutations, to ensure normal development. For example, several species may reproduce either asexually or sexually, but the adults are indistinguishable. However, just as he suggested, a lot of variability remained hidden in the genome. This means that development is not completely hardwired, in that, depending on environmental conditions, different pathways may be chosen that result in different phenotypes. More and more examples continue to be found that support Waddington's ideas. For instance, in some crocodiles and turtles sex determination is dependent on temperature, and in some fish it depends on the social context. Differences in gene expression can also cause differences in development. This is how the various castes in bees and ants are produced. Thus, if many genes are sensitive to environmental cues, it is not necessary to posit specific "plasticity" genes. Sometimes the differences might be a byproduct of environmental perturbations that were not protected by canalization. In other cases, it is a response to the environment and is clearly adaptive.

Organisms have evolved to use the environment as a source of important cues that can alter the trajectory of their development. Changes in gene expression may be as important as changes in gene function to promote plasticity. Environmental cues are often used to select the phenotype that appears most adaptive at that time. This facilitates a variety of evolutionary strategies. These include (1) **phenotypic accommodation** (adaptive change of variable aspects of the phenotype without genetic change due to novel inputs during development), (2) genetic assimilation (the process by which a phenotype is produced in response to something in the environment, such as exposure to a mutagen, and becomes encoded in the genome as a result of artificial or natural selection), and (3) **niche construction** (the process in which organisms significantly modify their own and sometimes others' selective environment such that it changes the selection pressures acting on present and future generations of the organism). An oft-cited example of phenotypic accommodation is the two-legged goat. Due to a birth defect a goat was born without functional forelimbs. By enlargement of the hind legs and changes in the spine and pelvis, it developed an upright posture and was able to walk on two legs. The classic example

If we return to the beginning of our story, we see that the most profound questions in trying to understand this astonishing phenomenon we call life were posed many centuries ago. But what is life? Even today we still do not fully understand how or when nonliving molecules arranged themselves in such a way that life emerged. However, an intrinsic property of something that was alive was that it was capable of reproducing. Thus, critical to understanding life was to unravel this fundamental problem of generation. Until the twentieth century, those early pioneers in the life sciences regarded the study of heredity and development not as separate, but merely two aspects of the same basic phenomenon, one that Aristotle had articulated in the fourth century BCE in his work *On the Generation of Animals*. It was more than 2,000 years before it was recognized that all living organisms were made up of cells, and that cells arise only from preexisting cells. That the cell is the smallest independent unit of life has shaped biological investigations ever since cell theory was proposed. One cannot underestimate the importance of the theory for biology but, like all good scientific theories, it should be able to accommodate new information and be modified if necessary.

The original ideas of both Schleiden and Schwann about how new cells were generated were disproven, but Virchow's dictum *omnis cellula e cellula* was correct. Nevertheless, the idea that the cell should be considered the smallest independent unit always had its detractors. Sachs proposed that the energide should be the smallest unit that was "alive." In particular, Thomas Huxley voiced his doubts and felt that the cell could not be considered an independent unit, not only the adult organism, but also most importantly as the embryo developed. For Huxley, the key to development would be found in understanding what was inside of the cell and how the whole organism did development, irrespective of cell boundaries. While some others were sympathetic to his ideas, they always represented a minority point of view throughout the twentieth century. Over the years new research programs emerged as embryologists investigated Aristotle's question of how form comes to be. At the same time the study of heredity became quite separate from embryology. However, an evolutionary approach is necessary to fully understand how an egg is awoken. Heredity and development must be brought back together. Development also needs to be more fully incorporated

of genetic assimilation are some experiments done by Waddington in the early 1940s. By subjecting fruit flies to ether vapor, he produced some mutant flies with a second thorax. By continuing to do this and selecting for the double thorax, he eventually was able to produce flies with a double thorax in the absence of the ether. The change had been incorporated into the developmental program of the fly.

Beavers exemplify the idea of niche construction. Beavers drastically shape and alter the ecosystem in which they live. Deforestation, effects on soil structure, root structure, turbidity of water, allocation of water, and the supply of water downstream are just a few the ways the beaver modifies its environment. In doing so it also affects the niche of many other organisms as well. The full evolutionary significance of organisms changing their environment, however, has only relatively recently begun to be studied. The term niche construction refers to such investigations.

Another evolutionary strategy that clearly has been highly successful is developmental symbiosis. In this case the developing animal utilizes cues from other organisms for normal cell differentiation and morphogenesis. Such symbiosis has been found to be ubiquitous. The coevolution of symbiotic microbes and animal cells has often led to an animal's developmental dependency on particular microbial signals, making these cues essential and expected components of normal development. Researchers are finding these signal factors are not generated only within the cell, but can come from sources external to the organism, and it is the environment that determines what structure will be made. In many species the genome has the ability to make a male or a female depending on a variety of factors. In addition to temperature, food supplies, pressure, and gravity, light, stress, and population density are all variables in particular species that determine whether the embryo develops as a male or a female. All of these processes emphasize the importance of nongenetic factors in shaping the course of both development and evolution. They also show that eco-devo is more than developmental plasticity, which focuses on the internal dynamics of how the genotype builds the phenotype.

What is the significance of these findings for cell theory? It challenges the idea that organisms are single genetic individuals and that the fundamental unit of an organism is the cell. Rather, as Gilbert and Epel suggest, organisms should be thought of more as ecosystems consisting of numerous genotypes interacting with each other: "This may allow natural selection to favor 'teams' rather than particular individuals and may also privilege 'relationships' as the unit of selection."[23] The whole notion of individuality has been fundamentally challenged. We have as many bacteria in our bodies as our "own" zygote-generated cells. The microbiome in our gut may help regulate the development of the brain and is turning out to be critical to our health. It influences our immune system and even our behaviors. Our symbiotic relationship with other organisms is critical at every phase of our existence, not just as an adult organism, but also in development. The truth of the matter is we have never been individuals. "We are all lichens."[24] None of this research undermines the fundamental importance of the cell in understanding what we call life. Just as the DNA dogma played a critical role in unraveling how the information coded in the gene impacts in very specific ways the activities of the cell, so has cell theory served as a foundation for furthering our understanding of organisms. Yet just as the DNA dogma has now needed to be modified, the time has come to realize that cell theory also needs to be revisited to further our understanding of development and the history of life.

23　Gilbert and Epel, *Ecological Developmental Biology,* chapter 11.
24　Gilbert, Sapp, and Tauber, "A Symbiotic View of Life," 341.

Epilogue

Omnia vivunt, omnia inter se conexa [Everything is everything is interconnected].

In considering the study of physical phenomen its noblest and most important result to be kno chain of connection, by which all natural force gether and made mutually dependent upon
Alexander vc

Our understanding of the cell has come a first peered through his microscope an cork as being made of cells. However, b how cells were thought of for hundre tural rather than the functional asp cell theory, it was recognized that living organisms was the cell. It br cell, particularly the nucleus. Th and the idea that the cell was important consequences fo standing development.

into evolutionary theory. Today, the importance of the environment in shaping both evolution and development is also being addressed. In doing so, aspects of cell theory that had always been problematic but had essentially been ignored are being revisited. Modern-day research is acknowledging the many valid points in Huxley's critique, even if various details of his description of cell structure were shown to be incorrect. All of this research also suggests that cell theory should be revised as we learn more and more about how a cell becomes a functioning organism in an ecosystem.

THE PROCESS OF SCIENCE

"Possible but not interesting. You'll reply that reality hasn't the least obligation to be interesting. And I'll answer you that reality may avoid that obligation, but that hypotheses may not."[1] So begins Daniel Mazia's 1987 paper on the centrosome cycle and the chromosome cycle. Progress in science has depended on interesting hypotheses. Many are disproven while others are regarded as improbable and are dismissed or ignored for generations. Huxley had several hypotheses that originally received short shrift but are now receiving the attention they deserve. His warning to Darwin, that he had burdened himself with gradualism and that the gaps in the fossil record were due to more than just its imperfection, received serious attention with the theory of punctuated equilibrium. "Punc eq" in turn paved the way for the concerns of developmental biologists that had been mainly passed over by the architects of the Modern Synthesis. Evo-devo addresses Huxley's primary interest in discovering the laws of form as evolutionary biologists realized that evolutionary theory was incomplete without incorporating development. This research along with that of eco-devo has also validated many aspects of Huxley's critique of cell theory. While biologists have recognized for a long time that the cell is dependent on other cells and environmental signals to function within an organism, very few have thought that should mean that cell theory should be altered. Yet evidence of many different kinds is accumulating that

---

1 Detective Lonnrot in Jorge Luis Borges's short story *Death and Compass*, quoted in Mazia, "The Chromosome Cycle and the Centrosome Cycle," 49.

suggests the cell should not be considered the smallest independent unit of life.

Mazia's career has also been characterized by interesting hypotheses that went against the prevailing ethos that favored the highly reductive approach of molecular biology with everything coming back to DNA. For Mazia, what could be more foundational to understanding life than unraveling what determines when a cell divides? He recognized the answer would not be found only in the genes. His provocative ideas about the dynamic nature of the centrosome resulted in him claiming that we had to fundamentally change how we thought about cells. Rather than giving primacy to the importance of the cell or the nucleus, he suggested that the basic unit should be the cell body, virtually identical to the concept of the energide, proposed more than 100 years earlier by Sachs. Mazia, however, had a lot more evidence to back up his claim, particularly his work in characterizing the mitotic apparatus and the role the centrosome plays in orchestrating the cell's activities. Much current research is bearing out his ideas.[2]

Much could be written about why ideas are accepted or not at a particular moment in history. Technological advances always play an important role. This history has pointed out many cases when a question couldn't even be investigated until the appropriate technology was invented or improved on. Technology also invites new questions. The microscope revealed a world of life that no one even thought existed. Improved microscopy also helped resolve various competing hypotheses about the structure and function of the cell and how development proceeded. New disciplines emerged and provided a wealth of new knowledge. At the same time, they shaped the kinds of questions that were investigated. Genetics was providing much more precise information about how the genes were transferred from one generation to another than did previous theories of heredity. This was aided by the fact that genetics was a much more narrowly defined discipline. Development had been removed from what had encompassed the previous meaning of heredity. Development became its own separate area of study, with its findings still largely descriptive, in spite of a more interventionist experimental approach toward the embryo. In evolution, population genetics with its mathematical modeling seemed much more precise than what paleontology had to offer in the 1930s and 1940s. Yet even then, George Gaylord Simpson recognized its limitations, writing that population genetics "may reveal what happened to one hundred rats in the course of ten years under fixed and simple conditions, but not what happened to one billion rats in the course of ten million years under the fluctuating conditions of earth history."[3] As biology went molecular, the dominance of genetics became even more pronounced. The study of development was its poor relation, in terms of both funding and prestige. No one disputes that the techniques of molecular biology have contributed enormously to our understanding of the cell. Yet at the same time, molecular genetics has contributed to the continued prominence of the highly reductive and gene-centric approach to understanding development. However, certain problems of biology have been resistant to these methods. That has never characterized our understanding of the cell, and Mazia never thought it should.

As important as advances in microscopy have been to our understanding of the cell, it must not be forgotten that what the microscope reveals is interpreted through the lens of theory and the eye of the observer. Pre cell theory, many people put forth various ideas about what was the basic structural unit of tissues based on what they "saw" through the microscope. We have seen how the personality of particular individuals influenced what interpretations were accepted or rejected. Due to the prominence of Schleiden and Schwann, other researchers interpreted their observations of cells dividing to make them compatible with the idea of cell-free formation. The same applies to the prestige of particular ideas. Ideas become deeply entrenched and are very difficult to displace. The main barrier to the widespread acceptance of the cell body/energide concept is resistance to the idea that the nucleus is a product of a symbiotic fusion event. This resistance is due to the predominant view that natural selection is the driving force of evolution, which minimizes the role symbiosis has played in the expansion and diversification of life. Thus, the acceptance of the cell body/energide as the smallest independent unit of life would not only mean a revision of cell theory, it indirectly challenges the dominant paradigm in evolution as well. Evolution by natural selection and cell

---

3 Simpson, *Tempo and Mode in Evolution*, xvii.

theory – there are not two more widely accepted ideas in all of biology! Any concept that suggests even a modification of these two core theories is going to have a very tough time indeed. Nevertheless, the history of science shows us time and time again that most scientific ideas are eventually shown to be wrong. One hundred years from now, will the cell body/energide be recognized as the smallest independent unit of life? Will cell theory become significantly modified to reflect both this idea and that the cell should not be considered as independent? Only time will tell.

However, we do not have to wait a hundred years to appreciate the importance of the questions that Huxley and Mazia raised. To fully understand how an egg develops into a cat or a redwood remains one of the great mysteries in biology and is intrinsically connected to solving one of the most difficult health problems of today. Cancer at its most fundamental level results when the normal regulation of the cell cycle is disrupted. For all the discussion of genes and cancer, a strictly genetic causation accounts for a relatively small percentage of cancers. The answer to the problem of cancer will be found when we can truly answer Huxley's question: how does form emerge? It emerges as cells divide and become specialized. Mazia devoted his career to trying to answer this most basic question: what makes a cell divide? And as he wrote more than 50 years ago in his investigations of mitosis, one can view the cell from the inside looking out, or from the outside looking in, but the answer will be found where these two views meet.

CONCLUSION

Cell biology is in transition from a science that was preoccupied with assigning functions to individual proteins or genes, to one that is now trying to cope with the complex sets of molecules that interact to form functional modules. The challenge for the twenty-first century is to understand how cells become organisms that survive and reproduce in an ever-changing environment. Although much of Mazia's work involved identifying molecules that were critical to the various stages of cell division, he always had this larger vision. He became more and more convinced that the structure of the cell played a profound role in the activities of the cell and its behavior. It is somewhat ironic that by privileging the structure of the cell, Hooke's description in many ways hindered the investigation of the functional aspect of what was occurring inside of the cell. In doing so, it set the stage for the cell being regarded as the smallest independent unit of life. Yet Mazia's interest in the structural aspects of the cell led him to his concept of the cell body and that the centrosome was the key to a deeper understanding of how one cell becomes two. This in turn had profound implications for the role of the centrosome in the developing embryo. He was a strong advocate for microscopy and believed that revealing the underlying structure of the cell, not only the centrosome, would provide insight to development. New methods of visualization have revolutionized our understanding of the cell. They have illuminated much about cellular dynamics, but at the same time raised complex issues about principles that underlie cell organization and morphogenesis.

Shaped by complementary reductionistic and holistic perspectives, Mazia's work was the embodiment of an organicist perspective. It also shows that as important as advances in technology are, theory and technology are inextricably linked. To the end of his life, Mazia was still engaged in research. He had hoped to participate in workshops in South Africa at the International Conference of Microscopists, but was unable to do so because of failing health. He sent his thoughts to the conference, which epitomize his thinking as a cell biologist:

> There are many paths in the advancement of science, but the giant leaps in our Science of the Cell have been made by seeing. First we see and then we interpret and only then do we pursue mechanisms and theories. The gifts of microscopes to our understanding of cells and organisms are so profound that one has to ask: What are the gifts of the microscopist? Here is my opinion. The gift of the great microscopist is the ability to THINK WITH THE EYES AND SEE WITH THE BRAIN. Deep revelations into the nature of living things continue to travel on beams of light.[4]

---

4 Epel and Schatten, "Daniel Mazia," 418.

On one of the last days Mazia spent in the lab, he modified these views slightly. After spending many hours peering into the microscope, he said to an assistant he would "see it when he believed it."[5] This is a reminder that what one sees is interpreted through both the lens of the microscope and the lens of the observer. The process of science is done by humans, and thus interpretation of data is subjective. This is not a criticism of science, but is essential to scientific progress. Mazia's comment does not apply just to microscopists, but to all scientists as they interpret experimental results. Great scientists "think with the eyes and see with the brain."

---

5 Epel and Schatten, "Daniel Mazia," 418.

Milestones and Controversies in the History of Cell Theory

PREVAILING THEORIES

Epigenesis 1700s–1800s

Theory first put forth by Aristotle that the embryo develops from unformed parts that gradually differentiate into tissue and organs, but also associated with vitalism and not widely accepted until the 1800s.

Preformation 1600s–1800s

The dominating theory that development occurred by the unfolding and growth of already fully formed structures.

Emboîtement Late 1600s–Early 1700s

Theory that the egg encases the germ of all future descendants that might develop from that egg, each germ encased within another germ. It contributed to the height of the popularity of the idea of preformation.

DRAMATIS PERSONAE IN THE HISTORY OF THE CELL[1]

Democritus 460–370 BCE

Greek philosopher who wrote that all matter is composed of minute, indivisible, and indestructible particles.

---

1 Some dates refer to the general period of time when the person was working, while specific dates refer to an important treatise that was published on that date.

Aristotle 384–322 BCE

Greek philosopher whose *On the Generation of Animals* is considered
to be the first scientific writing on embryology. It provided a comprehensive theory on generation and reproduction in a wide range
of animals and claimed that structures gradually formed out of
unformed matter or by epigenesis. He claimed that it was the male's
contribution that provided the "vital force" that animated lifeless
matter from the egg.

René Descartes 1644

French philosopher, scientist, and mathematician who argued that the
world was mechanistic, introducing the idea of materialism, which
stated that no vital force was necessary to animate life and contributed to the conceptual basis of preformation. Epigenesis was associated with vitalism at the time.

William Harvey 1651

English physician who is most famous for discovering the circulation of
the blood. Claimed that the egg, rather than the sperm, contained
the "vital force." The ova was the source of all life, originating in the
ovary of Eve. Ovism was the first conceptual model of preformation.

Robert Hooke 1665

English natural philosopher, polymath, and architect who investigated
many different fields in science and is credited with the discovery of
the cell. In his book *Micrographia* he described a variety of different
objects viewed under a microscope, including the structure of cork,
which he said was made up of cells, since they reminded him of the
structures found in honeycombs.

Francesco Redi 1668

Italian biologist, naturalist, and physician who challenged the theory of
spontaneous generation by demonstrating that maggots come from
the eggs of flies.

Anton van Leeuwenhoek 1671–1683

Dutch tradesman who had a distinguished second career as a scientist. He constructed his first simple microscope and continued to
improve the quality of the lenses in them, the best achieving a magnification of about 500 with a resolution of 1.0 μ. He also discovered

and described little "animalcules," including protozoa, amoebas, and bacteria. He described them as copulating, but what he probably observed was them simply dividing.

Jan Swammerdam 1637–1680

Dutch microscopist and biologist who confirmed the experiments of Redi. Considered the founder of the theory of preformation as a result of his research demonstrating that the rudiments of the adult structures such as legs and wings could be observed in the larval stages of various insects such as mayflies and butterflies.

Nicolas Malebranche 1670s

French philosopher and priest who claimed that the entire future of the human race was in the ovaries of Eve, like a series of nested Russian dolls.

Marcellus Malphigi 1675

Italian microscopist who in *Anatomes plantarum pars prima* described "utricles" (cells) as the basic unit of plant structure. Based on his microscopic observations, his detailed drawings of different stages of chicken embryos showed tiny, fully formed organs that needed only to unfold and grow to become a fully formed chick. This appeared to be strong evidence in favor of preformation.

Nehemiah Grew 1682

English plant anatomist and physiologist who described cell-like structures in his major treatise, *The Anatomy of Plants*. He could not find anything analogous to the veins and arteries observed in animals.

Nicolaas Hartsoeker 1694

Dutch mathematician and physicist who postulated that in the head of each sperm was a tiny homunculus or "little man."

Abraham Trembley 1744

Swiss naturalist who was the first to observe the division of several different species of newly discovered microscopic freshwater polyps. His research provided support for epigenesis.

John Needham 1745

English priest whose detailed experiments on various infusions seem to provide evidence in favor of spontaneous generation.

Lazzaro Spallanzani 1745

Italian scientist who, using a slightly different protocol than Needham, challenged Needham's findings regarding spontaneous generation. He also claimed to have observed binary fission in 14 different species of protozoa.

Albrecht von Haller 1757

Swiss animal physiologist whose investigations resulted in him claiming the basic structure of animal tissue were fibers. His views dominated the thinking about animal structure over those of Kaspar Wolff.

Kaspar Wolff 1759

German physician and biologist who in "Theoria Generationis" revived the theory of epigenesis. Based on his microscopic investigations, he argued that "globules" (he probably was observing cells) were the primary constituents of both plants and animal tissues and that fibers were secondary structures.

Felice Fontana 1781

Italian physicist who claimed that with the exception of certain membranes all animal tissue was made of twisted cylinders. Yet some of his own drawings clearly showed globules and to a modern eye look like flattened epithelial cells.

Marie François Xavier Bichat 1801

French anatomist and physiologist who is often referred to as the father of histology. He identified 21 different kinds of fibers that he claimed were the ultimate constituents of animal tissues.

Christian Pander 1817

German biologist and embryologist who identified three germ layers in the developing chick embryo.

Karl Ernst von Baer 1826

German biologist, naturalist, and embryologist who was the first person to have observed a mammalian egg in 1826. His investigations in comparative embryology resulted in his claim that animals could be grouped into four major types that all followed his four laws of development. His research provided compelling evidence in favor of epigenesis.

Johannes Müller 1830

German physiologist who made many contributions and who trained
many notable students, including Jacob Henle, Robert Remak, and
Rudolf Virchow. His assistant Theodor Schwann began his research
on the cell while working in Müller's lab.

Robert Brown 1831

British botanist who observed the nucleus in the cells of orchids. He
observed cytoplasmic streaming known as Brownian motion.

Félix Dujardin 1835

French cytologist who described material inside microscopic organisms
as a glutinous diaphanous substance that was insoluble in water and
contracted into a spherical mass that he called a sarcode.

Matthias Schleiden 1838

German botanist who is credited with the discovery that all plant parts
are made up of cells and with Theodor Schwann proposed cell theory in 1855. He renamed the nucleus the cytoblast and implied that
it was from here that new cells were generated *de novo*. Although he
was against the idea of preformation, he still thought the granules
that eventually formed the new cell were animated by some vitalistic
force.

Jan Purkyně 1839

A Czech anatomist and physiologist who confirmed the findings of
Dujardin about the substance inside the cell wall of plants, but
instead of a sarcode, called the material protoplasm. He discovered
large neurons in the cerebellum and was the first person to use a
microtome extensively in his research.

Albert von Kölliker 1840s

Swiss anatomist and physiologist who emphasized the critical importance the nucleus played in the life of the cell. Perhaps was the first
to observe and separate the mitochondria from other structures in
the cell (1888).

Jakob Henle 1841

German pathologist who described the elongation of nuclei that
then become long slender strands. It is not clear if he actually saw
chromosomes.

Carl Nägeli 1842

Swiss botanist who thought that cells divided by binary fission and in later research described what might have been chromosomes. At the same time he wrote that he had observed *de novo* synthesis of the nucleus, so it is apparent that he did not observe nuclear division.

Franz Unger 1844

Austrian biologist who described movement inside cells distinct from that of Brownian motion. He criticized Schleiden's theory of cell-free formation, describing the division of one cell into two.

Hugo von Mohl 1845

German botanist who described in detail the stages of mitosis in plant cells. He suggested the word "protoplasm" for the contents of the cell and that it was the source of the movements observed inside. His detailed observations of the structure of many different plants contributed to the overthrow of Schleiden's theory of cell-free formation.

Carl Zeiss 1846

Optician encouraged by Schleiden to found Zeiss Optical works in Jena, Germany, which became a leader in making high-quality microscopes.

Eduard Strasburger 1846

Polish-German cytologist who observed and described the discrete stages of mitosis that he named prophase, metaphase, and anaphase. In his investigation of uninuclear cells, he showed the process of cell division was homologous in plants and animals. In organisms with multinucleate cells, he emphasized the independence of cytokinesis (cell division) from mitosis.

Karl Reichart 1846

German anatomist, embryologist, and histologist who observed the furrowing of cells in the early stages of development. Instead of interpreting this as cell division, he described it in such a way as to be compatible with Schwann's theory of cell-free formation.

Theodor Schwann 1847

German physiologist who, along with Matthias Schleiden, proposed the cell theory. His work was primarily on animals, and he discovered giant cells of the nervous system called Schwann cells.

Wilhelm Hofmeister 1848

German botanist who described and illustrated in detail the stages
of mitosis that we know of today but referred to the chromatin as
"klumpen."

Ferdinand Cohn 1850

German botanist who was the first person to explicitly state that the sarcode of plants and the protoplasm of animals were virtually identical.

Robert Remak 1852

Jewish Polish-German embryologist and physiologist who first discovered that cells arise from preexisting cells, although Virchow is usually credited with the idea. His observations also disproved Schwann's
ideas of cell-free formation. He described and named the three layers of the early embryo: mesoderm, ectoderm, and endoderm. He
applied the word "protoplasm" to the yolk of the egg as well as to the
substance inside all embryonic animal cells.

Thomas Huxley 1853

British developmental morphologist who called himself "Darwin's bulldog," as he was an avid popularizer and defender of Darwin's theory.
He wrote an extensive review and critique of cell theory, claiming
the cell is not anatomically independent, but is dependent and
interacts with other cells in the organism. He had an epigenetic view
of development. In "The Physical Basis of Life" (1868) he popularized Max Schultze's idea that it is the contents or the protoplasm
inside the cell that is critical to understanding life, irrespective of
cell boundaries.

Franz Leydig 1857

German comparative anatomist and zoologist who helped the adoption
of the phrase "all cells arise from preexisting cells." He maintained
that the cell could be a morphologically complete unit even without
a cell wall.

Rudolf Virchow 1855

German physician, pathologist, and biologist who coined the phrase
"All cells arise from preexisting cells," the third postulate of cell theory and in opposition to Schleiden's and Schwann's theory of cell-
free formation.

Max Schultze 1861

German cytologist who defined the cell as a mass of protoplasm with a nucleus. He emphasized the contents of the cell rather than the structure of it as the critical aspect of the physical basis of life, which he called the protoplasmic theory of life.

Wilhelm His 1865

Swiss anatomist and cytologist who is sometimes credited with inventing the microtome. He distinguished and named the endothelium as separate from the epithelia membrane and clarified their relationship in the developing germ layers of the embryo. He advocated a research program in embryology that was more analytical and interventionist rather than primarily descriptive. He claimed that there were specific factors that caused differentiation localized in particular parts of the egg, but the egg did not contain preformed parts.

Louis Pasteur 1865

French microbiologist and chemist whose experiments demonstrated beyond reasonable doubt that living organisms only come from other living organisms, ruling out spontaneous generation and providing support for cell theory.

Ernst Haeckel 1866

German biologist who thought the nucleus contained the hereditary material of the cell. A proponent of evolution, he argued that ancestral stages of the adult were repeated in the embryonic stages of its descendants, known as Haeckel's biogenetic law: ontogeny recapitulates phylogeny. This contrasts with von Baer's views about how development occurred. He was the first person who really tried to integrate cell theory with evolutionary theory.

Oscar Hertwig 1871

German zoologist who observed that fertilization in sea urchins was the result of the fusion of the sperm with the nucleus of the egg. In 1876 he described the process of meiosis. He thought the hereditary material was located in the nucleus, but argued that cell differentiation was the result of the interaction between cells and was against the idea of mosaic development.

Hermann Fol 1876

Swiss biologist who confirmed Hertwig's experiments and definitively demonstrated fertilization in the starfish was the fusion of a single sperm with the ovum. However, continued research resulted in him claiming that the centrosome was a permanent structure exhibiting continuity generation to generation and was the key to understanding heredity rather than the nucleus.

Ernst Abbe 1878

German physicist and principal lens designer for Zeiss Optical Works. He designed an oil immersion lens in 1878 and in 1886 an apochromatic objective microscope that, in the hands of a skilled microscopist, made magnifications of 2500 diameters possible. This was almost the theoretical limit of what was possible with a compound microscope.

Charles Otis Whitman 1878, 1893

American biologist credited with the beginning of cell lineage studies that followed the fate of the cells in the developing embryo from the very first division of the fertilized egg. He concluded that development is determined, rather than regulative. The future of the organism is determined right from the initial organization of the egg. In addition, when and why a cell divided was determined by forces independent of cell boundaries, agreeing with Huxley's critique of cell theory.

Walther Flemming 1879–81

German biologist who thought the nucleus was the key to understanding heredity and described different stages of mitosis. He showed that the nuclear material took up stains and named it chromatin. He argued that the division of what later were called chromosomes was the key to division in all cells.

Edouard van Beneden 1883

Belgian cytologist and embryologist who, while working with the round worm *Ascaris*, identified definite corpuscles while observing mitosis. Boveri later named them centrosomes.

August Weismann 1885

German biologist who claimed that the hereditary factors were located in the germ plasm of the organism (which were eventually identified as chromosomes) and that the germ cells were sequestered from the rest of the body. This then precluded Lamarckian inheritance. He argued that development was mosaic, as a result of the hereditary determinants being divided unequally among the cells.

Theodor Boveri 1888

German biologist who discovered the organelle near the nucleus that he named the centrosome. He later wrote that it was the true division organ of the cell.

Julius von Sachs 1892

German botanist who criticized the adoption of the word "cell," claiming it set back the understanding of and investigations into life processes. Instead he proposed the concept of the energide as the primary constituent of all life.

Wilhelm Roux 1894

German zoologist who fully developed the research program in embryology that is known as *Entwicklungsmechanik*. As a result of his research on the development of frog embryos, he claimed that development was mosaic, that is, parts of the embryo developed independently of each other.

Hans Driesch 1880s–1900s

German embryologist who worked primarily on sea urchin development. Some experiments supported the findings of His and Roux that development was mosaic. However, many other experiments supported the idea that development was regulated by the organism as a whole. He also showed that cells of the early blastomere were "equipotent," that is, they had the ability to generate all the parts of the organism.

Jacques Loeb 1899

German-born American physiologist whose work on artificial parthenogenesis in invertebrates demonstrated that the sperm was not necessary to stimulate development. He had a fundamental disagreement with Boveri on the role sperm played in fertilization.

Edmund Beecher Wilson 1880s–1900s

American biologist who quickly accepted the chromosomal theory of inheritance and that the nucleus directed development. Yet he also wrote that the cell could not regarded as an isolated and independent unit. Rather, the only unity was of the entire organism.

Konstantin Mereschkowsky 1910

Russian biologist whose studies on lichens showed that, rather than a single organism, they consisted of a symbiotic partnership between an alga and fungus. He proposed a theory of symbiogenesis that claimed that complex large cells evolved from a symbiotic relationship of smaller ones.

Thomas Hunt Morgan 1890s–1900s

American geneticist and embryologist who is most famous for his work in genetics, particularly identifying hundreds of mutations in fruit flies. However, he started work as an embryologist and initially argued that it was the cytoplasm, rather than the nucleus and chromosomes, that directed development. He agreed with many of Huxley's criticisms of cell theory. He was one of the strongest advocates for separating development and heredity into two distinct disciplines, narrowly redefining heredity, which became the study of the transmission of genes. His early research on centophores supported the idea of mosaic development.

C.M. Child 1890s–1900s

American embryologist who, as a result of his research on regeneration, proposed the Gradient Theory in which he postulated the existence of specific physiological factors that guided regeneration and was convinced that they could be quantitatively measured. He thought the environment was responsible for inducing a metabolic gradient, which in turn provided the foundation for the physiological integration of the whole organism.

Frank R. Lillie 1900s

Canadian-born but one of the early pioneers of the American school of embryology, he demonstrated that normal development could occur quite far along, even in the absence of mitosis or cell division. He interpreted the findings to mean that cell division was not the primary factor in embryonic differentiation, but rather served as a process of localization. He also claimed that the whole organism was what determined how development occurred and that the cells were not independent.

E.G. Conklin 1900s

American embryologist who maintained that heredity was the problem
of differentiation over successive generations. He was an important
pioneer in cell lineage studies.

William Ritter 1919

American biologist who stressed the unity of life and proposed a theory
of living processes known as organicism. He was highly critical of cell
theory, emphasizing the organism as a whole. He objected to the separation of heredity and development into separate disciplines and was
critical of the more interventionist approach of studying organisms.

Electron Microscope 1931–present

Electron microscopes can magnify images thousands of times smaller
than a wave length of light, and they have been critical to furthering
understanding the fine structure of the cytoplasm.

Hans Spemann 1935

German embryologist who won the Nobel prize for his concept of
the organizer and embryonic induction, based on transplantation
experiments in amphibians. His research supported Huxley's idea
that the cell cannot be considered independent of the organism.

Sven Hörstadius 1939

Swedish embryologist who separated and recombined in different
ways the various layers of developing sea urchin embryos and then
followed the fate of particular cells. From these experiments he
hypothesized that the egg and early embryo contained particular
animalizing and vegetalizing substances that followed a concentration gradient, and that the ratio of these substances was responsible
for how development proceeded.

Paul Weiss 1930s–1950s

Pioneer in systems theory who thought that a deeper understanding
of morphogenetic fields would be the key to understanding development. He coined the phrase *Omnis organisatio ex organisatione.* All
organization from organization. He demonstrated that cells "know"
how to make the correct organ, even in an environment where they
aren't receiving any cues, and thought this was an even more fundamental process than induction.

Joseph Needham 1930s–1950s

English embryologist who, like Paul Weiss, advocated organicism and the importance of the morphogenetic field concept in trying to understand development. He maintained that the solution to the problem of pattern formation would only come about with the unification of biochemistry and morphogenesis and was influenced by C.M. Child's theory of metabolic gradients.

C.H. Waddington 1942

British biologist who did research to further clarify the concept of morphogenetic fields. He is best known for his concepts of canalization and the epigenetic landscape that attempted to bring heredity and development back together to understand morphogenesis.

Lynn Margulis 1967

American microbiologist who popularized and extended Konstantin Mereschkowsky's ideas of the importance of symbiosis as a major driving force in early evolution. She argued that the mitochondria and other organelles were once free living bacteria.

Daniel Mazia 1960s–1996

American cell biologist who, with Katsuma Dan, isolated and characterized the mitotic apparatus in sea urchin embryos. He eventually came to think that our view of the cell must be fundamentally changed. He proposed the concept of the cell body, which is virtually identical to Sach's concept of the energide. It is generated by a nuclear/centrosome complex that included the microtubule assemblage, condensing into the mitotic apparatus. He hypothesized that the centrosome might contain all the information about cell morphology and thus plays a critical role as to when a cell divides, which in turn will be crucial to understanding differentiation and development.

František Baluška 2004–present

Plant cell biologist who is calling for a revision of cell theory and arguing that the cell body/energide should be considered the smallest independent constituent of life. He proposes that the nucleus was the product of an endosymbiotic event.

Glossary

adaptation   (1) A character favored by natural selection because of its effectiveness in a particular role and that usually, but not always, contributes to reproductive success. (2) The process by which an organism adjusts to its environment.

adenine   One of four nitrogenous bases that appears in the DNA molecule and complementary base pairs with thymine.

alleles   Variants of a gene.

analogous   Referring to a similarity that is not due to common descent.

animal pole   The portion of an ovum that has less yolk, opposite the vegetable pole, and usually consists of smaller cells that divide rapidly as the embryo develops.

archaea   The group of single-celled organisms that are distinct from bacteria and eukaryotes and are one of the three major domains of life.

bacteria   Single-celled organisms that have cell walls but lack organelles and are one of the three major domains of life.

biodiversity   The variety of life on earth, but usually referring to the diversity found in a particular habitat or ecosystem.

blastema   A mass of cells capable of growth and regeneration into organs or body parts.

blastomeres   Cells that are the result of the first cell divisions or cleavage after fertilization where the large volume of the egg cytoplasm is divided into many smaller nucleated cells.

blastopore   The slit-like invagination of the gastrula through which cells move to form internal organs.

blastula   The early stage in the developing embryo consisting of a hollow cavity formed by a single layer of cells.

Brownian motion   The erratic random movement of microscopic particles in a fluid as a result of continuous bombardment from molecules of the surrounding medium.

canalization   The property of developmental systems to produce the same phenotype, in spite of environmental or genetic perturbations that might otherwise disrupt development.

cells   The basic structural building blocks of all organisms. The name is traced back to Robert Hooke's description of the structure of cork that he observed under the microscope.

cell body   A structure proposed by Daniel Mazia that was generated by the nuclear/centrosome complex that included the microtubular assemblage. Mazia hypothesized it was responsible for the organization of the entire intracellular structure.

cell cycle   The sequence of events that occurs in the cell that includes mitosis or cell division and interphase, which is involved in growth and/or preparation for division.

cell-free formation   The idea put forth by Schleiden and Schwann that new cells were formed *de novo* by a process of crystallization of cellular materials.

cell lineage   In development, refers to all the cells that can trace their origin back to one particular cell.

centrioles   Minute cylindrical organelles near the nucleus in animal cells, occurring in pairs and involved in the development of spindle fibers in cell division.

centromere   The region of a chromosome to which the microtubules of the spindle apparatus attach via the kinetochore during cell division.

centrosome   An organelle near the nucleus of a cell that contain the centrioles (in animal cells) and from which the spindle fibers develop in cell division. Theodor Boveri and Daniel Mazia thought it was the true division organ of the cell.

chromatin   The DNA and its associated proteins that form deeply stained material in the nucleus.

chromosomes   The rod-shaped elements that appear during cell division, consisting of one continuous DNA molecule and its associated proteins in eukaryotic cells. Also refers to the genetic material in prokaryotes.

codon   The sequence of three bases on the mRNA molecule that specifies a particular amino acid.

coenocytic   Containing many nuclei in one cell.

cytokinesis   The cytoplasmic division of the cell at the end of mitosis and meiosis.

cytoplasm   The jelly-like fluid inside the cell that includes all the structures except for the nucleus.

cytosine   One of four nitrogenous bases that appears in the DNA molecule and complementary base pairs with guanine.

de novo   To synthesize from scratch or anew.

determinant development   In development, referring to mitotic division of the fertilized ovum into blastomeres that are each destined to form a specific part of the embryo. Also known as mosaic development.

development   The process by which an embryo changes into a mature organism.

differentiation   The process by which cells become specialized and different from one another in the developing embryo.

diploid   Referring to a cell that has two copies of each chromosome.

DNA   The hereditary material deoxyribonucleic acid that consists of two long-chain molecules held together by hydrogen bonding. Each chain consists of a sequence of four bases: adenine (A), thymine (T), guanine (G), and cytosine (C).

dominant   Refers to an allele that affects the phenotype with only one copy.

dorsal   Referring to the upper side or back of an animal, plant, or organ.

ectoderm   In embryonic development, the outermost germ layer, whose cells become part of the nervous system, the sense organs, and the outer skin layer.

emboîtement   Theory promoted by the Catholic priest and philosopher Nicolas Malebranche that all living forms came from preexisting germs that were brought into being at the moment of creation. These encase the germs of all future living things, enclosed one within another. This theory contributed to the height of the popularity of the idea of preformation.

embryology   The study of the fertilization, early growth, and development of the egg of living organisms.

embryonic induction   The process by which one group of cells, the inducing tissue or organizer, directs the development of another group of neighboring cells, the responding tissue in the developing embryo.

endoderm   The innermost germ layer of cells or tissue that gives rise to the internal structures of the developing embryo, such as the lining of the digestive and respiratory systems.

endosymbiotic   Referring to the idea that the eukaryotic cell evolved from a symbiotic relationship of various prokaryotic cells and holds that various organelles such as the chloroplast and mitochondria were once free living. Not as well accepted is the idea that the origin of the nucleus is also a product of a fusion event.

energide   A structure, proposed by Julius von Sachs and later adopted by František Baluška, consisting of the nucleus and surrounding cytoplasm that should replace the cell as the fundamental unit of life.

entelechy   An immaterial force, proposed by Hans Driesch, that was what guided development toward completeness.

Entwicklungsmechanik   Developmental mechanics. A research program for studying development that was analytical and interventionist. It was first fully seen in the work of Wilhelm Roux.

epigenesis   A theory first put forth by Aristotle that the embryo develops from unformed parts that gradually differentiate into tissue and organs, but also associated with vitalism and not widely accepted until the 1800s.

epigenetic landscape   A concept representing embryonic development proposed by Conrad Waddington to illustrate the various developmental pathways a cell might take toward differentiation.

epigenetics   The study of inherited changes that does not involve a change in the actual DNA sequence, such as DNA methylation.

epimorphosis   A mechanism of regeneration that involves the dedifferentiation of adult structures to a mass of undifferentiated cells that then become respecified.

equipotent   Referring to cells in the developing embryo that have the ability to generate all parts of the organism.

eukaryotic   Referring to the group of organisms whose cells have a distinct nucleus and membrane-bound organelles.

evolution   Originally referred to progressive embryonic development, but now refers to Darwin's theory that organisms change through time, primarily by the mechanism of natural selection and resulting in a change in gene frequencies.

fertilization   A process in sexual reproduction that is the result of the fusion of the egg with the sperm.

gametes   The sex cells of an organism that contain half the number of chromosomes of regular cells.

gastrulation   The process of highly coordinated cell tissue movement in which the cells of the blastula are dramatically rearranged, giving rise to the three germ layers.

gene   The sequence of a DNA molecule that specifies the amino acid sequence of a polypeptide.

generation   Referring to procreation and includes both heredity and development in understanding how an organism "becomes."

genetic assimilation   The process by which a particular phenotypic character, which was initially produced in response to an environmental influence through a process of selection, then becomes canalized.

genotype   The genetic make-up of an individual.

germ plasm   The cells that are sequestered from the rest of the body destined to produce gametes. They contain the hereditary information that is transferred from generation to generation.

globules   Based on his microscopic observations, Kaspar Wolff claimed that globules were the primary constituents of both plants and animal tissues (he probably was observing cells).

gradients   Theory advanced by C.M. Child, who maintained that specific physiological factors were responsible for cell differentiation and that the environment induced a gradient of these factors.

guanine   One of four nitrogenous bases that appears in the DNA molecule and complementary base pairs with cytosine.

haploid   A cell that has only one copy of each gene.

heredity   The passing of genetic factors from parent to offspring or from one generation to the next.

heterozygous   Carrying two different alleles for the same gene.

histology   The microscopic study of the structure of tissues.

holism   The theory that because parts of a whole are interconnected, they cannot be understood without reference to the whole. The whole is greater than the sum of its parts, and a complete understanding cannot be reduced to simply understanding the parts.

homology   Anatomic structures or behavior traits in different organisms that are similar due to a shared common ancestor.

homozygous   Carrying identical alleles for the same gene.

homunculus   A fully formed "little man" that Nicholas Hartsoeker claimed existed inside the head of the sperm. This idea supported the idea of preformation.

infusoria   Referring to a collection of single-celled organisms such as ciliates, protozoa, unicellular algae, and small invertebrates that exist in freshwater ponds.

invertebrates   Animals lacking a backbone, such as arthropods, mollusks, annelids, and coelenterates.

in vitro   Performed or taking place in a test tube, culture dish, or elsewhere outside a living organism.

in vivo   Performed or taking place in a living organism.

kinetochore   A complex of proteins associated with the centromere of a chromosome during cell division, to which the microtubules of the spindle attach.

klumpen   The term Wilhelm Hofmeister used for chromatin in his detailed description and illustration of the stages of mitosis.

lichen   A complex life form that is a symbiotic partnership of a fungus and an alga.

mechanism   In philosophy, the belief that natural wholes (principally living things) are like complicated machines or artifacts and that a complete explanation of living organisms can be reduced to the laws of physics and chemistry.

meiosis   A type of cell division that gives rise to the sex cells and a halving of the genetic material.

mesoderm   The middle layer of an embryo in early development that is one of the three primary germ layers and gives rise to tissues including bone, muscle, and connective tissue.

microtome   A tool used to cut extremely thin slices of material, allowing for the preparation of samples for observation, particularly microscopic examination.

microtubule organizing center (MTOC)   The structure that appears in eukaryotic cells from which microtubules emerge. It organizes the mitotic and meiotic spindle apparatus that is responsible for separating the chromosomes during cell division.

microtubules   Microscopic hollow tubes made of the protein tubulin that are part of a cell's cytoskeleton and are a critical part of the mitotic apparatus that moves the chromosomes to opposite poles in mitosis.

mitosis In eukaryotic cells, the process of cell division that ensures each new cell has an exact copy of the genetic material.

mitotic apparatus A structure consisting of the centrosome and microtubules that aligns and separates the chromosomes during mitosis.

morphallaxis A mechanism of regeneration due to the repatterning of specific tissue owing to loss or death of the existing tissue and involves little additional cell growth.

morphogenetic field A group of cells able to respond to discrete, localized biochemical signals leading to the development of specific morphological structures or organs.

morphology The discipline in biology that deals with the form of living organisms and with relationships between their structures.

mosaic development Occurs when cell fate is determined early on in development and cells will develop into their predestined tissue even if transplanted to a new location in the embryo.

mutations Changes in the base sequence of the DNA molecule.

mutation theory A theory popular in the early 1900s that claimed that mutation, rather than natural selection, was the driving force of evolution.

neural plate A crucial structure that serves as the basis for the nervous system that forms from dorsal ectodermal tissue in the developing embryo.

niche construction The process in which organisms significantly modify their own and sometimes others' selective environment such that it changes the selection pressures acting on present and future generations of the organism.

notochord A semirigid rod running down the length of chordates (that includes all vertebrates and a few marine organisms such as tunicates).

nucleic acid Either DNA or RNA.

nuclein Substance isolated by Friedrich Miescher that contained DNA and protein.

nucleus The membrane organelle in the cell that contains the genetic material.

Omnis cellula e cellula All cells from cells. The phrase coined by Rudolf Virchow that became the third tenet of cell theory, that cells only arise from preexisting cells.

Omnis energide e energide   Phrase coined by František Baluška that means energides only arise from preexisting energides. The implication is that cells can only arise from other cells because the energide reproduces first.

Omnis organisatio ex organisatione   All organization from organization. The phrase coined by Paul Weiss to emphasize that the organism was not just the product of cells dividing, but was shaped by the organization of the whole developing embryo.

Omnis nucleus e nucleo   The idea put forth by Walther Flemming that all nuclei arise from nuclei.

ontogeny   The developmental history of an individual from the embryo to an adult.

organic   Referring to a class of chemicals that comes from once-living organisms, but now also refers to any carbon-based molecules.

organicism   The philosophy that living processes are a function of the entire, coordinated, autonomous system of an organism, rather than of any of its parts; the properties of the whole cannot be predicted solely from the properties of the component parts.

organizer   A special region of the embryo that is capable of determining the differentiation of other regions.

ovism   The idea put forth by William Harvey that the source of the life force existed in the ova rather than the sperm. Also the first conceptual model of the idea of preformation, that an organism is formed by the unfolding and growth of already preformed parts.

pangenesis   Heredity mechanism proposed by Darwin in which the cells throw off particles that collect in the reproductive organs so that the egg or bud contains particles from all parts of the parent.

parthenogenesis   Female reproduction in the absence of fertilization.

phenotype   The observable expression of an organism that is the product of the interaction between genes and the environment.

phenotypic accommodation   Adaptive change of variable aspects of the phenotype without genetic change due to novel inputs during development.

phylogeny   Referring to the evolutionary history of a species.

plasmodesmata   Connection between plant cells that allows the cytoplasm to flow between them.

pleiotropy   Referring to when one gene has multiple effects.

polyploid   A cell that contains extra sets of chromosomes.

preformation   The theory that development occurred by the unfolding and growth of already fully formed structures.

prokaryotic   Referring to single-celled organisms that do not have a distinct nucleus or other organelles; includes bacteria and archaea.

protein   A macromolecule that consists of long chains of amino acids and includes many important biological compounds such as enzymes, antibodies, and hormones.

protoplasm   Living matter, also defined as all the material inside of the cell.

protoplasmic theory   Theory proposed by Max Schultze in the nineteenth century. It argued that rather than the cell, what was inside the cell – the protoplasm – was what made something alive.

protozoa   Group that includes all single-celled eukaryotic organisms, either free-living or parasitic, that feed on organic matter including other microorganisms or tissues.

punctuated equilibrium   The theory claiming that the pattern of the fossil record indicates that most change occurs rapidly at the time of branching or speciation and then exhibits little or no change or stasis for long periods of time.

putrefaction   The rotting or decay of organic matter.

recapitulation   The theory that the development of an embryo passes through analogous phases as the species passes through its evolutionary or phylogenetic history.

recessive   An allele whose expression is masked by the activity of another allele.

reductionism   The philosophy that the properties of the whole can be known if all the properties of the parts are known.

regeneration   The reactivation of development in post-embryonic life to restore missing tissues.

regulative development   In early development, in which the cells are not committed to a definite fate and what they become is influenced by their position in the embryo.

RNA   Ribonucleic acid. A single-stranded nucleic acid that consists of nucleotides that contain ribose, phosphate, and the nitrogenous bases: guanine, uracil, cytosine, and adenine.

sarcode   The glutinous diaphanous substance inside microorganisms that Félix Dujardin described and later recognized to be the same as protoplasm.

slime mold   The group of organisms that exist as single ameboid cells but, depending on the species, will come together under suitable conditions and become a mass of creeping gelatinous protoplasm containing many nuclei or a mass of distinct cells.

spindle apparatus   The cytoskeletal structure in eukaryotic cells that forms during cell division to separate sister chromatids between daughter cells.

spontaneous generation   The belief that organisms can develop from nonliving matter.

supra-cellularity   Referring to a level of organization that is greater than the cell.

symbiogenesis   Theory proposed by Konstantin Mereschkowsky that more complex, large cells evolved from a symbiotic relationship of smaller ones.

symbiosis   When one organism lives on or with another organism, usually to the benefit of both.

thymine   One of four nitrogenous bases that appears in the DNA molecule and complementary base pairs with adenine.

totipotent   Referring to cells that are capable of giving rise to any cell type or from the blastomere developing into a complete embryo.

transplantation   In embryology an experimental technique where a portion of an embryo is replaced by a different portion from another embryo.

uracil   One of four nitrogenous bases that appears in the RNA molecule and complementary base pairs with adenine.

utricles   Marcellus Malphigi in 1675 described "utricles" (cells) as the basic unit of plant structure.

vegetable poles   The yolk-rich portion of ova that is opposite the animal poles. In the developing embryo they consist of larger cells that divide more slowly than cells in the animal poles.

ventral   Referring to the belly or underside.

vertebrates   Animals distinguished by the possession of a backbone or spinal column, including mammals, birds, reptiles, amphibians, and fishes.

vitalism   The idea that life is animated by some immaterial attribute or principle that is distinct from either purely chemical or physical forces.

zygote   The fused egg and sperm that develops into a diploid organism.

Bibliography

Adamatzky, Andrew, and Jeff Jones. "Road Planning with Slime Mold: If Physarum Built Motorways It Would Route M6/M74 Through Newcastle." *Nonlinear Sciences, Pattern Formation and Solitons* (2009). https://arxiv.org /abs/0912.3967.

Agnati, L.F., K. Fuxe, F. Baluška, and E.D. Guidolin. "Implications of the 'Energide' Concept for Communication and Information Handling in the Central Nervous System." *Journal of Neural Transmission* (February 2009). https://doi.org/10.1007/s00702-009-0193-1.

Allchin, Douglas. "Error Types." *Perspectives on Science* 9 (2001): 38–59.

———. *Sacred Bovines*. Oxford: Oxford University Press, 2017.

Allen, Garland. "Mechanism, Vitalism and Organicism in Late Nineteenth and Twentieth-Century Biology: The Importance of Historical Context." *Studies in History and Philosophy of Biological and Biomedical Science* 36 (2005): 261–83.

Amundson, Ron. *The Changing Role of the Embryo in Evolutionary Thought*. New York: Cambridge University Press, 2007.

Avery, Oswald T., Colin M. Macleod, and Maclyn McCarty. "Studies on the Chemical Nature of the Substance Inducing Transformation of Pneumococcal Types: Induction of Transformation by a Desoxyribonucleic Acid Fraction Isolated from Pneumococcus Type III." *Journal of Experimental Medicine* 79, no. 2 (December 1944): 137–58.

Baer, Karl Ernst von. "Philosophical Fragments *Uber Entwickelungsgeschichte The Fifth Scholium*." 1828. In *Scientific Memoirs, Natural History*, edited by A. Henfrey and T. Huxley. London: Taylor and Francis, 1853.

Baluška, František. "Cell-Cell Channels, Viruses, and Evolution Via Infection, Parasitism, and Symbiosis: Toward Higher Levels of Biological Complexity." *Annals of the New York Academy of Sciences* 1178 (2009): 106–19.

————. *Communication in Plants.* Berlin: Springer, 2006.

Baluška, František, P.W. Barlow, I.K. Lichtscheidl, and D. Volkmann. "The Plant Cell Body: A Cytoskeletal Tool for Cellular Development and Morphogenesis." *Protoplasma* 202 (1998): 1–10.

Baluška, František, Andrej Hlavacka, Dieter Volkmann, and Diedrik Menzel. "Getting Connected: Actin-Based Cell-to-Cell Channels in Plants and Animals." *Trends in Cell Biology* 14 (2004): 404–08.

Baluška, František, and Sherrie Lyons. "Symbiotic Origin of Eukaryotic Nucleus: From Cell Body to Neo-Energide." In *Concepts in Cell Biology History and Evolution,* edited by V.P. Sahi and F. Baluška, 39–66. Berlin: Springer Verlag, 2018.

Baluška, František, and Velemir Ninkovic, eds. *Plant Communication from an Ecological Perspective.* Berlin: Springer-Verlag, 2010.

Baluška, František, Dieter Volkmann, and Peter W. Barlow. "Actin-Based Domains of the 'Cell Periphery Complex' and Their Associations with Polarized 'Cell Bodies' in Higher Plants." *Plant Biology* 2 (2000): 253–67.

————. "Cell Bodies in a Cage." *Nature* 428 (2004): 371.

————. "Cell-Cell Channels and Their Implications for Cell Theory." In *Cell-Cell Channels,* edited by František Baluška, Dieter Volkmann, and Peter W. Barlow, 1–17. New York: Springer Science + Business Media-2006.

————. "Eukaryotic Cells and Their Cell Bodies: Cell Theory Revised." *Annals of Botany* 94 (2004): 9–32.

————. "Nuclear Components with Microtubule Organizing Properties in Multi-cellular Eukaryotes: Function and Evolutionary Considerations." *International Review of Cytology* 175 (1997): 91–135.

Baluška, František, Dieter Volkmann, Diedrik Menzel, and Peter W. Barlow. "Strasburger's Legacy to Mitosis and Cytokinesis and Its Relevance for the Cell Theory." *Protoplasma* 249, no. 4 (2012): 1151–62. https://doi .org/10.1007/s00709-012-0404-8.

Barnett, Heather. The Slime Mould Collective (website). https://slimoco .ning.com/profile/HeatherBarnett.

Bauer, Henry. *Scientific Literacy and the Myth of the Scientific Method.* Chicago: University of Illinois Press, 1992.

Beloussov, Lev V., with additional commentary by John Opitz and Scott Gilbert. "Life of Alexander G. Gurwitsch and His Relevant Contribution to the Theory of Morphogenetic Fields." *International Journal of Developmental Biology* 41 (1997): 771–79.

Blinderman, Charles, and David Joyce. The Huxley File (website). https:// mathcs.clarku.edu/huxley/.

Boveri, Theodor. "Heft 4. Ueber die Natur der Centrosomen." *Zellen-Studien.* Jena, Germany: Fischer, 1901.

Bowen, James. "The Scientific Revolution of the 17th Century." In *The Coral Reef Era: From Discovery to Decline,* edited by J. Bowen, 11–19. Dordrecht: Springer Science+Business Media, 2015. https://doi.org 10.1007/978-3-319-07479-5.

Brachet, Jean, and Alfred E. Mirsky, eds. *The Cell: Biochemistry, Physiology, Morphology*, vol. 3. New York: Academic Press, 1961.

Bradbury, S. *The Microscope Past and Present*. Oxford: Pergamon Press, 1968.

Brenner, Eric D., Rainer Stahlberg, Stefano Mancuso, Jorge Vivanco, František Baluška, and Elizabeth Van Volkenburgh. "Plant Neurobiology: An Integrated View of Plant Signaling." *Trends in Plant Science* 8 (11 August 2006): 413–19. https://doi.org/10.1016/j.tplants.2006.06.009.

Carroll, Sean. *Endless Forms Most Beautiful: The New Science of Evo Devo*, reprint ed. New York: W.W. Norton & Company, 2006.

Caspersson, T., E. Hammarsten, and H. Hammarsten. "Interactions of Proteins and Nucleic Acids." *Transactions of the Faraday Society* 31 (1935): 367–89.

Chisholm, A.D. *Cell Lineage*. New York: Academic Press, 2001. https://doi.org/10.1006/rwgn.2001.0172.

Claude, Albert. "Nobel Lecture: The Coming Age of the Cell." https://www.nobelprize.org/prizes/medicine/1974/claude/lecture/.

Coen, Enrico. *Cells to Civilization*. Princeton: Princeton University Press, 2012.

Cohn, Ferdinand. "Zur Naturgeschichte des Protococcus pluvialis." *Nova Acta Academiae Caesareae Leopoldino-Carolinae* 22 (1850): 605–764.

Coleman, William. "Cell, Nucleus, and Inheritance: An Historical Study." *Proceedings of the American Philosophical Society* 109, no. 3 (15 June 1965): 124–58.

Conklin, Edward Grant. "The Mechanism of Heredity." *Science* 27 (1908): 89–99.

Cossins, Dan. "Plant Talk." *The Scientist* 1 (2014): 37–43.

Crick, Francis. *What Mad Pursuit*. New York: Basic Books, 1988.

Crow, Ernst W., and James F. Crow. "100 Years Ago: Walter Sutton and the Chromosome Theory of Heredity." *Genetics* 160, no. 1 (1 January 2002): 1–4. http://www.genetics.org/content/160/1/1.full.

Darlington, Cyril. "Interaction Between Cell Nucleus and Cytoplasm." *Nature* 140 (1937): 932.

Darwin, Charles. *The Correspondence of Charles Darwin 1856–1857*, edited by Frederick Burkhardt and Sydney Smith. Cambridge: Cambridge University Press, 1985. https://www.darwinproject.ac.uk.

———. *Life and Letters of Charles Darwin*, 2 vols. Edited by Francis Darwin. London: John Murray, 1887.

———. *More Letters of Charles Darwin*, 2 vols. Edited by Francis Darwin and A.C. Seward. London: John Murray, 1903.

———. *On the Origin of Species*, 1st ed. London: John Murray, 1859. Reprint, New York: Avenel, 1976.

Di Trocchio, Federico. "Mendel's Experiments: A Reinterpretation." *Journal of the History of Biology* 24, no. 3 (1991): 485–519.

Dobell, Clifford. *Antony van Leeuwenhoek and His "Little Animals"; Being Some Account of the Father of Protozoology and Bacteriology and His Multifarious Discoveries in These Disciplines*. New York: Staples Press, 1932.

Dobzhansky, Theodosius. *Genetics and the Origin of Species*, 2nd ed. New York: Columbia University Press, 1941.

———. "Nothing in Biology Makes Sense Except in the Light of Evolution." *American Biology Teacher* 35, no. 3 (March 1973): 125–29.

Doxsey, Stephen, Wendy Zimmerman, and Keith Mikule. "Centrosome Control of the Cell Cycle." *Trends in Cell Biology* 6 (15 June 2005): 303–11.

Eldredge, Niles, and Stephen J. Gould. "Punctuated Equilibria: An Alternative to Phyletic Gradualism." In *Models in Paleobiology*, edited by T.J.M. Schopf, 82–115. San Francisco: Freeman Cooper, 1972.

Elwick, James. *Styles of Reasoning in the British Life Sciences: Shared Assumptions, 1820–1858.* London: Pickering and Chatto, 2007.

Embryo Project Encyclopedia. https://embryo.asu.edu/info/embryo-project.

Epel, David, and Gerald Schatten. "Daniel Mazia: A Passion for Understanding How Cells Reproduce." *Trends in Cell Biology* 8, no. 10 (1 October 1998): 416–18.

Esposito, Maurizo. "More than the Parts: W.E. Ritter, the Scripps Marine Association, and the Organismal Conception of Life." *Historical Studies in the Natural Sciences* 45, no. 2 (April 2015): 273–302.

———. *Romantic Biology 1890–1945.* London: Pickering & Chatto, 2013.

Evered, David, and Joan Marsh, eds. *Cellular Basis of Morphogenesis.* Ciba Foundation Symposium 144. New York: John Wiley & Sons, 1989.

Flannery, Maura C. "Images of the Cell in Twentieth-Century Art and Science." *Leonardo* 31, no. 3 (1998): 195–204.

Friedman, William, and Pamela K. Diggle. "Charles Darwin and the Origins of Plant Evolutionary Developmental Biology." *Plant Cell* 23, no. 4 (April 2011): 1194–207.

Garcia-Bellido, Antonio. "Cellular Interphase." In *Cellular Basis of Morphogenesis*, edited by David Evered and Joan Marsh, 8. Ciba Foundation Symposium 144. New York: John Wiley & Sons, 1989.

Geison, Gerald. "The Protoplasmic Theory of Life and the Vitalist-Mechanist Debate." *Isis* 60, no. 3 (1969): 272–92.

Gershon, Nahum D., Keith R. Porter, and Mark A. McNiven. "Three Dimensional Structure of the Cell Center Revealed by Computer Graphics Methodology." *Biophysics Journal* 49 (1986): 65–66.

Gieryn, Thomas. *Cultural Boundaries of Science.* Chicago: University of Chicago Press, 1999.

Gilbert, Scott F. "The American Precursors of Evo-Devo: Ecology, Cell-lineages, and Pastimes Unworthy of the Deity." *Theory in Biosciences* 127, no. 4 (2008): 291–96.

———. "Bearing Crosses: The Historiography of Genetics and Embryology." *American Journal of Medical Genetics* 76 (1998): 168–82.

———. "Cells in Search of Community: Critiques of Weismannism and Selectable Units in Ontogeny." *Biology and Philosophy* 7 (1992): 473–87.

———. "Cellular Politics: Just, Goldschmidt, and the Attempts to Reconcile Embryology and Genetics." In *The American Development of Biology*, edited by R. Rainger, K. Benson, and J. Maienschein, 311–46. Philadelphia: University of Pennsylvania Press, 1988.

———. "Conceptual Breakthroughs in Developmental Biology." *Journal of Biosciences* 23 (1998): 169–76.

———. "Cytoplasmic Action in Development." *Quarterly Review of Biology* 66 (1991): 309–31.

———. "Ecological Developmental Biology: Developmental Biology Meets the Real World." *Developmental Biology* 233 (2001): 1–12.

———. "The Embryological Origins of the Gene Theory." *Journal of the History of Biology* 11 (1978): 307–51.

———. "Epigenetic Landscaping: Waddington's Use of Cell Fate Bifurcation Diagrams." *Biology and Philosophy* 6 (1991): 135–54.

———. "In Friendly Disagreement: Wilson, Morgan, and the Embryological Origins of the Gene Theory." *American Zoologist* 27 (1982): 797–806.

———. "Induction and the Origins of Developmental Genetics." In *A Conceptual History of Modern Embryology*, edited by Scott Gilbert, 181–206. New York: Plenum Press, 1991.

———. "Intellectual Traditions in the Life Sciences: Molecular Biology and Biochemistry." *Perspectives in Biology and Medicine* 26 (1982): 151–62.

———. "The Morphogenesis of Evolutionary Developmental Biology." *International Journal of Developmental Biology* 47 (2003): 467–77.

———. "The Role of Embryonic Induction in Creating Self." In *Organism and the Origins of Self*, edited by A.I. Tauber, 341–60. Dordrecht: Kluwer Press, 1991.

———. "Symbiosis as a Way of Life: The Dependent Co-origination of the Body." *Journal of Biosciences* 39 (2014): 201–09.

Gilbert, Scott F., and J. Atkinson, eds. "Development and Macroevolution." *American Zoologist* 32 (1992): 101–44.

Gilbert, Scott F., and Jonathan Bard. "Formalizing Theories of Development: A Fugue on the Orderliness of Change." In *Towards A Theory Of Development*, edited by Alessandro Minelli and Thomas Pradeu, 129–43. Oxford: Oxford University Press, 2014.

Gilbert, Scott F., Thomas C.G. Bosch, and Cristina Ledón-Rettig. "Eco-Evo-Devo: Developmental Symbiosis and Developmental Plasticity as Evolutionary Agents." *Nature Reviews Genetics* 169 (2015): 611–22.

Gilbert, Scott F., and Richard M. Burian, "Development, Evolution, and Evolutionary Developmental Biology." In *Key Words and Concepts in Evolutionary Developmental Biology*, edited by Brian Hall and Wendy Olson, 68–74. Cambridge, MA: Harvard University Press, 2003.

Gilbert, Scott, F., & David Epel. *Ecological Developmental Biology*. New York: Oxford University Press, 2009.

Gilbert, Scott F., John M. Opitz, and Rudolf A. Raff. "Resynthesizing Evolutionary and Developmental Biology." *Developmental Biology* 173 (1996): 357–72.

Gilbert, Scott F., Jan Sapp, and Alfred I. Tauber. "A Symbiotic View of Life:
    We Have Never Been Individuals." *Quarterly Review of Biology* 87 (2012):
    325–41.
Gilbert, Scott F., and Sahotra Sarkar. "Embracing Complexity: Organicism for
    the 21st Century." *Developmental Dynamics* 219 (2000): 1–9.
Gilbert, Scott F., Sahotra Sarkar, and Alfred I. Tauber. "Symposium on the
    Evolution of Individuality: Introduction." *Biology and Philosophy* 7 (1992):
    461–62.
Gilbert, Scott F., and Lauri Saxé. "Spemann's Organizer: Models and
    Molecules." *Mechanisms of Development* 41 (1993): 73–89.
Goodwin, Brian. *How the Leopard Changed Its Spots: The Evolution of Complexity.*
    New York: Charles Scribner's Sons, 1994.
Gould, Stephen J., and Niles Eldredge. "Punctuated Equilibria: The Tempo
    and Mode of Evolution Reconsidered." *Paleobiology* 3, no. 2 (1977): 115–51.
Grew, Nehemiah. *The Anatomy of Plants.* London: W. Rawlins, 1682.
Grzybowski, Andrzej, and Pietrzak Krzysztof. "Robert Remak (1815–1865)."
    Journal of Neurology 260, no. 6 (June 2013): 1696–97. https://doi.org
    /10.1007/s00415-012-6761-6.
Hall, Brian. "Evolutionary Developmental Biology (Evo-Devo): Past, Present,
    and Future." *Evolution and Education and Outreach* 5 (2012): 184–93.
———. *Homology: The Hierarchical Basis of Comparative Biology.* New York:
    Academic Press, 1994.
———. "Ontogeny Does Not Recapitulate Phylogeny, It Creates Phylogeny."
    Evolution and. Development 132 (2011): 401–04.
———. "Parallelism, Deep Homology, and Evo-Devo." *Evolution and
    Development* 14 (2012): 29–33.
———. "Waddington's Legacy in Development and Evolution." *American
    Zoologist* 32 (1992): 113–22.
Haller, Albrecht von. *Elementa physiologiae corporis humani.* Lausanne: M.M.
    Bousquet, 1757.
Haraway, Donna. *Crystals, Fabrics, and Fields: Metaphors That Shape Embryos.* New
    Haven: Yale University Press, 1976.
Harris, Henry. *The Birth of the Cell.* London: Yale University Press, 1999.
Hershey, A.D., and Martha Chase. "Independent Functions of Viral Protein
    and Nucleic Acid in Growth of Bacteriophage." *Journal of General Physiology*
    36 (1952): 39–56.
Hinchcliffe, Edward H., Frederick, J. Miller, Matthew Cham, Alexey
    Khodjakov, and Greenfield Sluder. "Requirement of a Centrosomal Activity
    for Cell Cycle Progression through G_1 into S Phase." *Science* 291, no. 5508
    (23 February 2001): 1547–50.
Hooke, Robert. *Micrographia,* 1665. Lincolnwood, IL: Science Heritage,
    facsim. ed., 1987.
Hopkins, Sir Frederick Gowland. "Some Aspects of Biochemistry." *Irish Journal
    of Medical Science* 7, no. 7 (1932): 333–50.

Huxley, Julian. *Evolution: The Modern Synthesis.* New York: George Allen and
    Unwin, 1942.
Huxley, Leonard, ed. *Life and Letters of Thomas Henry Huxley,* 2 vols. New York:
    D. Appleton.
Huxley, Thomas Henry. "Agnosticism." 1889. In *Science and the Christian
    Tradition,* 209–62. London: Macmillan, 1894.
———. "Autobiography." In *Method and Results,* 7. New York: D. Appleton, 1897.
———. "The Cell Theory." *British and Foreign Medico-Chirurgical Review* 12
    (1853): 285–314. In *Scientific Memoirs of Thomas Henry Huxley,* Vol. 1,
    edited by M. Foster and E. Ray Lancaster, 242–78. London: Macmillan,
    1898–1902.
———. "Criticisms on the Origin of Species." 1864. In *Darwiniana,* 80–106.
    London: Macmillan, 1893.
———. "Evolution in Biology." 1878. In *Science and Culture,* 207–309. London:
    Macmillan, 1881.
———. "Further Evidence of the Affinity Between the Dinosaurian Reptiles
    and Birds." In *Scientific Memoirs of Thomas Henry Huxley,* Vol. 3, edited by
    M. Foster and E. Ray Lancaster. London: Macmillan, 1898–1902. First
    published 1870.
———. "The Genealogy of Animals." 1869. In *Darwiniana,* 107–19. London:
    Macmillan, 1893.
———. "Letter no. 2119." Darwin Correspondence Project. http://www
    .darwinproject.ac.uk/DCP-LETT-2119.
———. "On the Anatomy and the Affinities of the Family of the Medusae." In
    Scientific Memoirs of Thomas Henry Huxley, Vol. 1, edited by M. Foster and E.
    Ray Lancaster. London: Macmillan, 1898–1902. First published 1849.
———. "On the Animals Which are Most Nearly Intermediate Between Birds
    and Reptiles." In *Scientific Memoirs of Thomas Henry Huxley,* Vol. 3, edited
    by M. Foster and E. Ray Lancaster. London: Macmillan, 1898–1902. First
    published 1868.
———. "On the Border Territory Between the Animal and the Vegetable
    Kingdoms." In *Scientific Memoirs of Thomas Henry Huxley,* Vol. 4, edited by
    M. Foster and E. Ray Lancaster. London: Macmillan, 1898–1902. First
    published 1876.
———. "On the Identity of the Structure of Plants and Animals." In *Scientific
    Memoirs of Thomas Henry Huxley,* Vol. 1, edited by M. Foster and E. Ray
    Lancaster. London: Macmillan, 1898–1902. First published 1853.
———. "On the Physical Basis of Life." 1869. In *Method and Results,* 130–65.
    New York: D. Appleton, 1897.
———. "The Reception of the 'Origin of Species'." 1887. In *Life and Letters of
    Charles Darwin,* 2 vols., edited by Francis Darwin, vol. 1, 533–58. New York:
    D. Appleton, 1900.
———. *Scientific Memoirs of Thomas Henry Huxley,* 4 vols., edited by M. Foster
    and E. Ray Lancaster. London: Macmillan, 1898–1902.

———. "Some Considerations upon the Meaning of the Terms Analogy and Affinity." Thomas Henry Huxley Collection, 37.1. Imperial College of Science and Technology London.

———. "The Study of Zoology." 1861. In *Lay Sermons, Addresses and Reviews*, 83. London: Macmillan, 1899.

———. "Upon Animal Individuality." *Edinburgh New Philosophical Journal* 53 (1852): 172–77.

Jabr, Ferris. "How Brainless Slime Molds Redefine Intelligence." *Scientific American* (7 November 2012). https://www.scientificamerican.com /article/brainless-slime-molds.

Jacyna, L.S. "The Romantic Programme and the Reception of Cell Theory in Britain." *Journal of the History of Biology* 17, no. 1 (Spring 1984): 13–48.

Judson, Horace Freeman. *The Eighth Day of Creation: Makers of the Revolution in Biology*. New York: Simon and Schuster, 1979.

Koestler, Arthur, and J.R. Smythies, eds. *Beyond Reductionism*. London: Hutchinson, 1969.

Laubichler, Manfred D., and Jane Maienschein, eds. *From Embryology to Evo-Devo: A History of Developmental Evolution*. Cambridge, MA: MIT Press, 2007.

Lewis, Ricki. *Human Genetics, The Basics*. New York: Routledge, 2011.

Lidgard, Scott, and Lynn K. Nyhart, eds. *Biological Individuality*. Chicago: University of Chicago Press, 2017.

Lillie, Frank R. "Observations and Experiments Concerning the Elementary Phenomena of Embryonic Development in *Chaetopterus*." *Journal of Experimental Zoology* 3 (1906): 153–268.

Liu, Daniel. "The Cell and Protoplasm as Container, Object, and Substance, 1835–1861." *Journal of the History of Biology* 50 (2017): 889–925. https://doi .org/10.1007/s10739-016-9460-9.

Loeb, Jacques. *The Mechanistic Conception of Life*. Chicago: University of Chicago Press, 1912.

———. *The Organism as a Whole: From a Physicochemical Viewpoint*. New York: G.P. Putnam's Sons, 1916.

Lyons, Sherrie. *Evolution: The Basics*. New York: Routledge Press, 2011.

———. *Species, Spirits, Serpents, and Skulls: Science at the Margins in the Victorian Age*. New York: SUNY Press, 2009.

———. *Thomas Henry Huxley: The Evolution of a Scientist*. New York: Prometheus Books, 1999.

Maienschein, Jane. "Cell Theory and Development," in *Companion to the History of Modern Science*, edited by R.C. Olby, G.N. Cantor, J.R. Christie, and M.J.S. Hodge, 357–73. London: Routledge, 1990.

———. *Defining Biology Lectures from the 1890s*. Cambridge, MA: Harvard University Press, 1986.

———. *Transforming Traditions in American Developmental Biology*. New York: Johns Hopkins University Press, 1991.

Malphigi, Marcello. *Anatomes Plantarum.* London, 1675 and 1679.

Margulis, Lynn. *Origin of Eukaryotic Cells.* New Haven: Yale University Press, 1970.

———. "Serial Endosymbiotic Theory (SET) and Composite Individuality: Transition from Bacterial to Eukaryotic Genomes." *Microbiology Today* 31 (2004): 172–74.

———. *Symbiosis in Cell Evolution.* San Francisco: W.H. Freeman, 1993.

Matlin, Karl S., Jane Maienschein, and Manfred D. Laubichler, eds. *Visions of Cell Biology: Reflections Inspired by Cowdry's General Cytology.* Chicago: University of Chicago Press, 2018.

Mayr, Ernst. *Systematics and the Origin of Species from the Viewpoint of a Zoologist.* New York: Columbia University Press, 1942.

Mazia, Daniel. "The Cell Cycle at the Cellular Level." *European Journal of Cell Biology* 61, Suppl. 38 (1993): 14.

———. "Centrosomes and Mitotic Poles." *Experimental Cell Research* 153, no. 1 (1984): 1–15.

———. "The Chromosome Cycle and the Centrosome Cycle in the Mitotic Cycle." *International Review of Cytology* 100 (1987): 49–92.

———. "Mitosis and the Physiology of Cell Division." In *The Cell: Biochemistry, Physiology, Morphology,* vol. 3, edited by Jean Brachet and Alfred E. Mirsky, 77–412. New York: Academic Press, 1961.

———. "Mitotic Poles in Artificial Parthenogenesis: A Letter to Katsuma Dan." *Zoological Science* 5 (1988): 519–28.

———. "Origin of Twoness." In *Cell Reproduction: In Honor of Daniel Mazia,* edited by Ellen R. Dirksen, David M. Prescott, and C. Fred Fox, 1–14. New York: Academic Press, 1978.

———. "Physiology of the Cell Nucleus." In *Modern Trends in Physiology and Biochemistry,* edited by E.G.S. Barron, 77–122. New York: Woods Holes Lectures, 1952.

Mazia, Daniel, and Katsuma Dan. "The Isolation and Biochemical Characterization of the Mitotic Apparatus of Dividing Cells." *Proceedings of the National Academy of Sciences of the United States of America* 38 (1952): 826–38.

Mazzarello, Paolo. "A Unifying Concept: The History of Cell Theory." *Nature Cell Biology* 1, no. 1 (May 1999): E13–15. https://doi.org/10.1038/8964.

McShea, Daniel W. "Complexity and Homoplasy." In *Homoplasy: The Recurrence of Similarity in Evolution,* edited by M.J. Sanderson and L. Hufford, 207–27. San Diego: Academic Press, 1996.

Meijering, Erik. "Cell Segmentation: 50 Years Down the Road." *IEEE Signal Processing Magazine* 29, no. 5 (September 2012): 140–45. https://doi.org/10.1109/MSP.2012.2204190.

Merton, Robert. "The Matthew Effect in Science." *Science New Series* 159, no. 3810 (1968): 56–63.

Minelli, Alessandro, and Thomas Pradeu, eds. *Towards a Theory of Development.* Oxford: Oxford University Press, 2014.

Mitman, Gregg, and Anne Fausto-Sterling. "Whatever Happened to Planaria? C.M. Child and the Physiology of Inheritance." In *The Right Tool for the Job: At Work in Twentieth-Century Life Sciences*, edited by A.E. Clarke and J.H. Fujimura, 172–97. Princeton: Princeton University Press, 1992.

Moore, John. A. *Science as a Way of Knowing: The Foundations of Modern Biology*. Cambridge, MA: Harvard University Press, 1993.

Morgan, Thomas Hunt. "Chromosomes and Heredity." *American Naturalist* 44 (1910): 449–96.

———. *The Frog's Egg*. New York: MacMillan, 1897.

———. "The Relation of Genetics to Physiology and Medicine." 1934. In *Nobel Lectures in Physiology and Medicine 1922–1941*. Amsterdam: Elsevier, 1967.

———. *The Theory of the Gene*. New Haven, CT: Yale University Press, 1926.

Morgan, Thomas Hunt, A.H. Sturtevant, H.J. Muller, and C.B. Bridges. *The Mechanism of Mendelian Heredity*. New York: Henry Holt, 1915.

Moritz, Karl, and Helmut Sauer. "Boveri's Contributions to Developmental Biology: A Challenge for Today." *International Journal of Developmental Biology* 40 (1996): 27–47.

Mukherjee, Siddhartha. "Soon We'll Cure Diseases with a Cell, Not a Pill." Ted Talk, 6 October 2015. https://www.ted.com/talks siddhartha_mukherjee.

Needham, Joseph. *Chemical Embryology*. London: Cambridge University Press, 1931. Reprint, New York: Hafner, 1963.

Nicholson, Daniel J. "Biological Atomism and Cell Theory." *Studies in History and Philosophy of Biological and Biomedical Sciences* 41 (2010): 202–11.

Nurse, Paul. "The Incredible Life and Times of Biological Cells." *Science* 289, no. 5485 (8 September 2000): 1711–16.

Olby, Robert. "Mendel, No Mendelian?" *History of Science* 17 (1979): 53–72.

———. *The Path to the Double Helix*. New York: Dover, 1974, 1994.

Oppenheimer, Jane. *Essays in the History of Embryology and Biology*. Cambridge, MA: MIT Press, 1967.

Ospovat, Dov. "The Influence of Karl Ernst von Baer's Embryology, 1828–1859: A Reappraisal in Light of Richard Owen and William B. Carpenter's Paleontological Application of von Baer's Law." *Journal of the History of Biology* 9 (1976): 1–28.

Owen, Richard. "Report on the Archetype and Homologies of the Vertebrate Skeleton." British Association of the Advancement of Science, 1846.

Pasquero, Claudia, and Marco Poletto. "Cities as Biological Computers." *Architectural Research Quarterly* 20, no. 1 (2016): 10–19. https://doi.org/10.1017/S135913551600018X.

Provine, William. "Progress in Evolution and Meaning of Life." In *Evolutionary Progress*, edited by M. Nitecki, 49–74. Chicago: University of Chicago Press, 1989.

Radick, Gregory. "Teach Students the Biology of Their Times." *Nature* 533, no. 7603 (19 May 2016): 293.

Reynolds, Andrew S. "Discovering the Ties That Bind: Cell-Cell
    Communication and the Development of Cell Sociology." In *Biological
    Individuality*, edited by S. Lidgard and Lynn K. Nyhart, chapter 4. Chicago:
    University of Chicago Press, 2017.
——. "The Theory of the Cell State and the Question of Cell Autonomy in
    Nineteenth and Early Twentieth-Century Biology." *Science in Context* 20,
    no. 3 (2007): 71–95.
——. *The Third Lens: Metaphor and the Creation of Modern Cell Biology.*
    Chicago: University of Chicago Press, 2018.
Richmond, Marsha. "T.H. Huxley's Criticism of German Cell Theory: An
    Epigenetic and Physiological Interpretation of Cell Structure." *Journal of the
    History of Biology* 33 (2000): 247–89.
Rieder, Conly L., and Alexey Khodjakov. "Mitosis Through the Microscope:
    Advances in Seeing Inside Live Dividing Cells." *Science* 300, no. 5616
    (4 April 2003): 91–96. https://doi.org/ 10.1126/science.1082177.
Ritter, William. *The Unity of the Organism.* Boston: R.G. Badger, 1919.
Ritterbush, Philip. *The Art of Organic Forms.* Washington, DC: Smithsonian
    Institute Press, 1968.
Rudwick, Martin. *The Meaning of Fossils.* Chicago: University of Chicago Press,
    1976.
Russell, E.S. *Form and Function: A Contribution to the History of Animal
    Morphology*, 1916. Chicago: University of Chicago Press, 1992.
Sapp, Jan. *Evolution by Association: A History of Symbiosis.* Oxford: Oxford
    University Press, 1994.
——. "Freewheeling Centrioles." *History and Philosophy of the Life Sciences* 20
    (1998): 255–90.
——. *The New Foundations of Evolution.* Oxford: Oxford University Press,
    2009.
Schatten, Gerald, and Tim Stearns. "Sperm Centrosomes: Kiss Your Asterless
    Goodbye, for Fertility's Sake." *Current Biology* 25 (21 December 2015):
    R1178–81. http://dx.doi.org/10.1016/j.cub.2015.11.015.
Schatten, Heide, Gerald Schatten, Daniel Mazia, Ron Balczon, and Calvin
    Simerly. "Behavior of Centrosomes During Fertilization and Cell
    Division in Mouse Oocytes and in Sea Urchin Eggs." *Proceedings of the
    National Academy of Sciences of the United States of America* 83, no. 1 (1986):
    105–09.
Schultze, Max. *Archiv für Naturgeschichte* 26 (1860): 287–310.
Schwann, Theodor. *Microscopical Researches into the Accordance in the Structure
    and the Growth of Animals and Plants*, translated by Henry Smith. London:
    Sydenham Society, 1847.
Scott, Michon. "Robert Hooke." Strange Science: The Rocky Road to Modern
    Paleontology and Biology (website). www.strangescience.net/hooke.htm.
Shadwell, Thomas. *The Virtuoso*, 1676. Reprint, Lincoln: University of
    Nebraska Press, 1966.

Simpson, George Gaylord. *Tempo and Mode in Evolution.* New York: Columbia University Press, 1944.

Smith, John Maynard. "In Conversation with John Maynard Smith." 2 February 1999. www.academia.edu/6101586/In_conversation_with _John_Maynard_Smith_FRS_1999_.

Stanford News. "Cell Biologist Daniel Mazia Dies at 83." 11 June 1986. https:// news.stanford.edu/pr/96/960611mazia.html.

Stebbins, G. Ledyard. *Variation and Evolution in Plants.* New York: Columbia University Press, 1950.

Stern, William T. "Robert Brown." 2008. Encyclopedia.com: http://www .encyclopedia.com/doc/1G2-2830900668.html.

Studnicka, F.K. *Acta Soc. Scient. Natural Moravicae* 4 (1927): fasc: 4: 1.

Sunderland, Mary E. "The Gradient Theory." Embryo Project Encyclopedia (website). http://embryo.asu.edu/handle/10776/1689.

Sutton, Walter S. "On the Morphology of the Chromosome Group in *Brachystola magna.*" *Biological Bulletin* 4 (1902): 24–39.

Thompson, D'Arcy. *On Growth and Form.* Cambridge: Cambridge University Press, 1917.

Tournier, F., and Michel Bornens. "Centrosome and Cell Cycle Control." In *Microtubules*, vol. 13, edited by Jeremy S. Hyams and Clive W. Lloyd. New York: Wiley-Liss, 1994.

Van Valen, Leigh. "Festschrift." *Science* 180 (1973): 488.

Vertii, Anastassia, Heidi Hehnly, and Stephen Doxsey. "The Centrosome: A Multitalented Renaissance Organelle." *Cold Spring Harbor Perspectives in Biology.* New York: Cold Spring Harbor Laboratory Press, 2016. https://doi .org/10.1101/cshperspect.a025049.

Virchow, Rudolf. *Archiv für Anatomie, Physiologie und Wissenschaftliche Medicin* 11 (1857).

———. *Die Cellular-Pathologie*, 3rd ed. Berlin: Verlag von August Hirschwald, 1862.

Waddington, Conrad. *The Evolution of an Evolutionist.* New York: Cornell University Press, 1975.

Walczak, Claire E., Shang Cai, and Alexey Khodjakov. "Mechanisms of Chromosome Behaviour During Mitosis." *Nature Reviews Molecular Cell Biology* 2 (11 February 2010): 91–102. https://doi.org/10.1038/nrm2832.

Watase, Shosaburo. *Biological Lectures Delivered at the Marine Biological Laboratory of Woods Hole in the Summer Session of 1894.* Boston: Ginn and Co., 1895.

Waters, C. Kenneth, and Albert Van Helden, eds. *Julian Huxley: Biologist and Statesman of Science*, 1927. Reprint, Texas: A & M University Press, 2010.

Watson, James D., and Francis Crick. "Molecular Structure of Nucleic Acids: A Structure for Deoxyribose Nucleic Acid." *Nature* 171 (1953): 737–38.

Webster, Gerry, and Brian Goodwin. *Form and Transformation.* Cambridge: Cambridge University Press, 1996.

Weiss, Paul. "Macromolecular Fabrics and Patterns." *Journal of Cellular Comparative Physiology* 49, Supp. 1 (1957): 105–12.

———. *Principles of Development*. New York: Holt, Rinehart, and Winston, 1939. Reprint, New York: Haftner, 1969.

———. "The Problem of Cell Individuality in Development." *American Naturalist* 74 (1940): 34–46.

Weiss, Paul, and Aron Moscana. "Type-Specific Morphogenesis of Cartilages Developed from Dissociated Limb and Scleral Mesenchyme *in vitro*." *Journal of Embryological Experimental Morphology* 6 (1958): 238–46.

Welch, G. Rickey, and James S. Clegg. "From Protoplasmic Theory to Cellular Systems Biology: A 150-Year Reflection." *American Journal of Physiology and Cell Physiology* 298, no. 6 (2010): C1280–90.

West-Eberhart, Mary Jane. "Biography." *Evolution and Development* 11, no. 1 (2009): 8–10. https://doi/org/10.1111/j.1525-142X.2008.00297.x.

———. *Developmental Plasticity and Evolution*. New York: Oxford University Press, 2003.

Whitman, Charles Otis. "The Embryology of Clepsine." *Quarterly Review of Microscopical Science* 18 (1878): 215–315.

———. "The Inadequacy of the Cell-Theory in Development." *Journal of Morphology* 8, no. 3 (1893): 639–58. https://doi.org/10.1002/jmor.1050080307.

Wilson, Edmund Beecher. "Amphioxus and the Mosaic Theory of Development." *Journal of Morphology* 8 (1893): 579–639.

———. *The Cell in Development and Inheritance*. New York: Macmillan, 1896, 1900, 1912, 1925.

———. "The Mosaic Theory of Development." *Biological Lectures Delivered at the Marine Biological Laboratory*, vol. 2, 1–14. Woods Hole: Marine Biological Laboratory, 1893.

Wilson, Edmund Beecher, and Edward Leaming. *An Atlas of the Fertilization and Karyokinesis of the Ovum*. New York: Macmillan, 1895.

Wilson, Leonard, ed. *Sir Charles Lyell's Scientific Journals on the Species Question*. New Haven: Yale University Press, 1970.

Woese, Carl R. "A New Biology for a New Century." *Microbiology and Molecular Biology Reviews* 68 (2004): 173–86.

———. "On the Evolution of Cells." *Proceedings of the National Academy of Sciences of the United States of America* 99 (2002): 8742–47.

Wolff, Kaspar, 1759. *Theoria generationis, edit nova, aucta et emendata*, 2nd ed. Christ. Hendel, Halae ad Salam, 1774.

Wolpert, Lewis. "Evolution of Cell Theory." *Philosophical Transactions of the Royal Society B* 329, no. 1329 (September 1995): 227–35.

Yandell, Kate. "Sketching Out Cell Theory, circa 1837." *The Scientist* 27, no. 8 (August 2013): 72.

Index

Page numbers in italics refer to figures.